CW00408103

UFOs D(T

The Greatest Lie that Enveloped the World

Dr. Roger Leir

Foreword by Apollo 14 Astronaut
Edgar Mitchell, Ph.D.

THE BOOK TREE
San Diego, California

© 2014 Roger Leir

All rights reserved

No part of this publication may be used or transmitted in any way without the expressed written consent of the publisher, except for short excerpts for use in reviews.

ISBN 978-1-58509-142-3

Front cover design
A. J. Gevaerd

Cover layout
Mike Sparrow

Interior layout & design
Paul Tice

Published by
The Book Tree
P O Box 16476
San Diego, CA 92176

www.thebooktree.com

We provide fascinating and educational products to help awaken the public to new ideas and information that would not be available otherwise.
Call 1 (800) 700-8733 for our FREE BOOK TREE CATALOG.

CONTENTS

FOREWORD

BY EDGAR MITCHELL, Ph.D.

Mitchell on the moon, NASA

In the twentieth century the invention of aircraft and then spacecraft created new approaches to the ancient questions about the basic nature of the universe that we inhabit. With new technologies emerging from those discoveries of aviation and spaceflight, and other new inventions allowing us to look ever more deeply into the heavens, we discover a myriad of new galaxies, each with billions of star systems, and the total array extending into the heavens seemingly to infinity. On that evidence alone, the question: "Are we alone in the universe?" is the equivalent of suggesting that one particular grain of sand on a beach is markedly different than all the other grains of sand on all the beaches of the world.

In addition, however, the evidence that we have been visited by beings from some of those distant worlds is becoming overwhelmingly obvious to those who look carefully at the evidence. Whether the visitations are increasing in frequency is an open question. Some evidence suggests that visitation has been occurring for hundreds of years. However, our science and its evidence can only postulate such occurrences; there is little hard evidence to point toward ancient visitation.

5

The famous incident termed the "Roswell Incident" in 1947 near Roswell, New Mexico, has become a landmark event in the study of UFOs and caused intensified questions about alien presence and intentions. In the more than six decades since that event, the study of UFO sightings, alien presence and activities has consumed the attention of well-educated and qualified investigators. In spite of denials and serious efforts by many governments to deny, the evidence has continued to mount for alien presence and activities. The learned and intelligent citizenry is becoming convinced that, indeed, our Earth is being visited by an advanced technological species. As our space faring technology on Earth has not yet provided humans the capability to explore extensively within our own solar system, with manned vehicles, much less go beyond, it is obvious that we humans have much to learn about exploring more deeply into Nature.

Some of the most interesting series of reports of alien activity are the reports by military personnel of interference by the visitors with military rockets capable of conducting intercontinental warfare. Many such rockets, ready to launch from their silos toward a prospective enemy, have been disabled by UFOs. The evidence, though hidden from public disclosure at the time, has been available to serious researchers. It is good to know that the visitors do not approve of our preparations for a global war. It certainly makes the alien presence less threatening. This is not to imply that all the events with aliens have been reported as peaceful and benign. There have been cases of reported abductions, and testing of human specimens against their will. The available literature has numerous events of both types. It sounds not too different than we humans going into new territories and uncharted waters and being less than cooperative with the natives.

The explosion of human population during the 20th century from less than 2 billion persons to over 7 billion in less than 100 years, and still continuing to grown in the 21st century, with the attendant consumption of limited resources and increased technological creation that strains Earth's capability to sustain life, is a major cause of concern to those who look at the larger perspective. It is most likely that the alien visitors are aware of the limits that evolution places on planets in a solar system. Their presence is not without knowledge of Earth's predicament of population and limiting resources. We can hope that at the proper time, they may be helpful in solving this very difficult problem.

As we humans have now begun our own travels in space, we may be better able to anticipate and prepare for the required technologies and the new understanding that a modern perspective of reality requires. The universe (or, multi-verse, if you prefer) is vastly larger and more complex than our ancestors, even early in the 20th century, could possibly imagine. Our trips to our moon, and the reality of experiencing another heavenly body has begun the process of reaching outward toward the distance stars – a process that will surely accelerate with time and as new knowledge comes to be.

My own experience of visiting the moon on Apollo 14 allowed an explorer's impulses to be vented. The press of time and limited resources caused us to follow our planned and practiced regimen as precisely as possible. And the practice on Earth to be timely and precise paid off. We were able to do our planned exploration and collect samples of the moon just as had been planned. Of course the reduced gravity and the lack of atmosphere could not be easily practiced on Earth, but the anticipation allowed adequate accommodation with that reality. As explorers bent on collecting samples and data from our designated landing site we were quite successful in accomplishing those tasks; thus proving the point that study and prolonged practice pay off in success of the actual venture.

Edgar Mitchell

The following is a series of questions posed to Dr. Mitchell by the author pertaining to his personal experiences on the moon.

1. What were your thoughts when you actually placed your foot on another planetary body?

After taking in and absorbing the new landscape, the thoughts were to get back to work, and do the tasks we had practiced for this landing site.

2. What was it like to stand for the first time and gaze over the close and distant landscape of the moon? Was this comparable to any place you had been on Earth?

No, the desolate, dry, desert like crater-marked surface without any atmosphere and with 1/6 the gravitation has no real Earthly comparison.

3. Were you able to see the stars when you looked overhead?

No, for the same reason you cannot see stars during the day from the Earth's surface: the Sun is too bright and obscures starlight. If you were to adequately shield you eyes and sun reflections you could see the stars, as there is no atmosphere to scatter the star light.

4. How far did you actually travel when you traversed the surface of the moon?

Our geology target was Cone Crater, a 1000-foot across and 800-foot deep impact crater. It was 2.5 kilometers from our landing point. We walked to it and back, collecting soil and rock samples as they represented the soil and rock thrown out from inside the crater when it formed centuries ago.

5. Where there any strange reflections in your face plate, since there was no atmosphere to look through?

No.

6. What were your personal thoughts when you saw the Earth from your position on the moon?

It was small and bright, and directly overhead. It was quite a sight to realize Earth was 240,000 miles away.

7. Did you notice at any time distortions in any portion of the moon's environment that might have been caused by other than natural or environmental causes?

No.

8. How did shadows appear on the moon without the refraction of an atmospheric lens?

Just as dark areas devoid of light.

9. Did you have thoughts that any living creatures had been on the moon other than humans?

No.

10. Did you notice at any time distortions in soil that might have been caused by other than natural or environmental causes?

No.

INTRODUCTION

The sky was a brilliant blue and it was much to early for stars to be seen. Suddenly three bright stars appeared and slowly approached the couple. They had no fear and were in awe as the craft came closer and closer. Both were fascinated by their beauty.

The following is based on a true story. The names and places have been altered for privacy purposes.

Phil and Virginia were in their teen years, Phil being 18 years old and Virginia only 17.

Their feelings for each other could not be described in any terms other than sublime love. We, as adults, might call those feelings infatuation but nevertheless, their bond was mutual and strong. Phil might be described as a rather conservative young man, about 5 feet 8 inches tall and rather lanky in build. His main interest in high school was running track. He had mastered numerous track and field events and received the appropriate awards for his achievements. Although he was never much of a social enthusiast, he was well respected by his friends and peers. Phil did quite well academically and maintained about a B average for his efforts. In contrast Virginia, nicknamed Ginny, was what many might consider the ultimate in beauty. She stood about 5 foot 4 inches in height with dark brown hair which flowed like a magnificent waterfall, cascading to her mid back. She was blessed with an "hourglass figure," which most of the time was hidden by her conservative attire. Ginny's personality was as far removed from that of her boyfriend as it could possibly be. Phil was raised in upstate New York, and Virginia was from the deep South of Alabama. Her nature was gentile and soft. She was also very spiritual due to her family upbringing. Together they made a strange combination but nevertheless, their emotions thrived.

It was a magnificent spring day and they had decided to trek through some property owned by the Department of Forestry. Carried with them was the makings for a picnic lunch, replete with a bottle of non-alcoholic champagne. The spring sun bore down on their faces and bathed them in the glorious Southern California spring sun. There was a magical reddish sheen, which glistened in the appropriate lighting. A gentle breeze tingled Ginny's sparkling hair, blowing wisps of her crowning glory in her eyes. She constantly swept her hand over her face to keep these menacing hair strands from tickling her. Each time she uttered a tiny laugh, which naturally gained Phil's attention.

It was now about one o-clock in the afternoon when they found their ideal spot to set out their picnic essentials. This spot was shielded by the shade of an isolated and typically twisted oak tree. The grass and ground around them was ideal for the placement of their blanket.

Ginny wasted no time in gathering her typical southern menu, complete with fried chicken, crisp fried okra and homemade bread. Phil opened the pseudo champagne bottle and poured two glasses. They made a toast to each other with their arms entwined, each letting the other sip from the glass held by the other. Phil stood up and grasped Ginny by the hand, helping her rise to her feet. He gently put his arm around her waist and pulled her closer to him. They stood there staring into each other eyes with an overwhelming look of passion and affection. Slowly their lips met. Time stopped for both of them as if neither wanted the moment to end. Suddenly Ginny opened her eyes, and forcibly shoved Phil away from her. Phil was flabbergasted at this negative gesture from his beloved Ginny. With that, she stepped back and said in a surprised be serene voice, "Phil, look over your right shoulder and tell me what you see." He turned quickly in that direction. They stood looking at what appeared to be three bright stars in the bright blue cloudless sky. They were amazed at what they were viewing and watched as the stars got brighter and closer. Phil reached for Ginny in a protective stance but neither felt any fear. Closer and closer they came. Brighter and larger were the lights until they could see the outline of the light's origin. They were three metallic discs. Both Phil and Ginny gazed skyward in amazement as the ships approached them at a low altitude. They were beautiful, giving off an array of colors and maneuvering in various geometric patterns that to them seemed to be putting on their own private show. Neither Ginny or Phil experienced any fear or trepidation but instead felt as if they knew who was operating the craft. Their adventure lasted for at least five minutes and suddenly the craft shot off into the vast sky in the blink of an eye. Phil was the first to speak. "Ginny, do you have any idea of what we just witnessed?" Before Phil could speak again, Ginny asked, "Did they talk to you too?" Phil's jaw dropped and remarked surprisingly, "Do you mean you heard them also? I thought I was going nuts and was not going to tell you." He grabbed her by the shoulders and asked, " What did they say to you?" Ginny looked into Phil's eyes and told him that she was told they were her space brothers and sisters and there was no need to fear them. She also told him that someday they would return for the purpose of preserving mankind.

Phil and Virginia have been married for years and remember this occurrence as if it had just happened yesterday. They are married with five children and many grandchildren.

It should be noted that they never referred to these craft as UFO's. They told me they knew exactly what they were and described them as being three non-terrestrial spacecraft. Based on similar sightings throughout the world, the big question has always been, "Do UFOs exist?" UFOs may be real, based on thousands of legitimate sightings, but are not "factually" real according to so-called experts. Why doesn't the term "exist" have any basis in fact? Perhaps what personalizes a "fact" is one's concept of reality. Perhaps reality is individualized to each human. Our concepts of our environs are in a continuous state of flux and for one instant (as, for example, in the case of a UFO sighting) our perception may be dominated by forces outside our accepted concept of reality. Then there is an opposing view, which suggests our reality is controlled and molded by forces which we accept in our everyday life. It is not the purpose of this author to bore the reader with a mass of dictionary meanings of the term, but to familiarize the reader with the exact meanings of words we have been using during the modern era. Surprisingly enough, our language and word meanings may help with the entire problem of the UFO dilemma and the perceptions we share of them. Dictionary examples will be listed in the Appendix and represent the exact meaning of the words which have been in use for a prolonged period of time.

Perhaps at one time, because of numerous political, military and social reasons, certain truths were kept from the public with genuine concern for the citizenry and general population of the world. There might have been genuine justification for withholding the truth on the basis of religious grounds alone. Many scholars have said if we have no knowledge or understanding of the past, then how can we understand and even begin to predict the future? I believe the same parallel exists between the genuine meaning of words, their understanding and continued understanding of the truth.

We, as human beings, do not understand our origin or what we are. We have been kept in complete ignorance about our origin as well biological and spiritual make up. How was this accomplished? It is this author's belief that it was done through our ability to manipulate the understanding of our language – and to use language itself, which is based on individual words. The information contained in the last Appendix will help close the bond between the true meaning of the "WORD" and how it relates to the subject of UFO's. Please refer to the last Appendix for this precise information, but at the same time always remember the case of Phil and Virginia.

CHAPTER 1
THE ADVENTURE BEGINS

It has been said that to understand the future, one must know the past. I have accepted this as a fact of life and made a sincere endeavor to correlate past occurrences in the field of Ufology with my predictions of the future. At the time when I became a serious researcher in this very strange field of study, there were certain subjects that were popular with the large active research groups such as CUFOS *(Center for UFO Studies)*, N.I.C.A.P. *(National Investigations Committee on Aerial Phenomena)*, and MUFON *(Mutual UFO Network)*; yet, while there were broad areas of acceptance of some of the basic interests of the subject, others were purposely neglected. Most interest was centered on eye-witness testimonies and available photos taken with the old film cameras. Honor was also given to accepted pieces of physical evidence such as landing traces and other unexplained disturbances of the intimate environment where an unknown flying object was seen. If individuals reported that the engine of their automobile died, then that became an investigative fact. Early investigative organizations were often under the severe scrutiny of propaganda mills that were able to twist the facts to suit their own uses. Some resolved only to investigate the possibility that vehicles of unknown origin were invading our skies, but gave little attention to the fact that they were seemingly intelligently controlled. The research took a giant leap forward when attention was focused on this possibility of intelligent control. The next logical step was to look into WHO WAS CONTROLLING THEM. Such serious discussions led to the "Extraterrestrial Hypothesis." Occasionally individuals claimed to have had direct contact with the pilots of these vehicles: Frank Critzer and George Van Tassel were two such persons. Their experiences inspired them to create an airport for these space vehicles in Giant Rock, California. Their interests gave way to much publicity in the 1950s—and some today say that their experiences were credible.

Next came an era when not only the unidentified vehicles were deemed okay to study, but also their pilots, with some accepting the possibility that these "unknown beings" could have come from another planet in our solar system. After all, canals had been spotted on Mars for many years, with some of the popular pseudo-scientific belief that intelligent life did indeed dwell on that planet.

Almost immediately upon the heel of this emerging acceptance came the amazing "kidnapping" of two people in 1961. Yes! The abduction case of Betty and Barney Hill occurred in New Hampshire in September of that year. The Hill case underwent as much investigation as any well-known murder case on record, yet the verdict remained in limbo for many years. There are those researchers who cast their doubts even today, despite the research gained from the Hills' experience which has resulted in major contributions to the depth of our understanding. Betty Hill's recollection of a different sort of "Star Map" which she had been shown while onboard the alien craft was one such benefit. With her account, and the work of amateur astronomer Marjorie E. Fish, we learned of and have come to accept a twin star system known as Zeta Reticuli.

I believe the Hill abduction case opened the door for acceptance of further alien abduction cases such as the 1973 Pascagoula, Mississippi incident involving Calvin Parker and Charles Hickson who were both taken by alien beings aboard a craft against their will. This was followed by the November 5, 1975 case of lumberer Travis Walton who was taken from a forested area of Arizona. Travis's case was different in that, at the time of his abduction, he had no conscious memory of being abducted, but stated that he was standing in close proximity to the craft as it was powering up. At that instant, a bolt of energy shot out of the vehicle, hitting him squarely in the chest and knocking him some distance from where he had been standing.

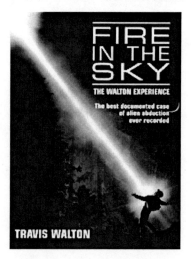

According to the police department and his close colleagues who were present at the time of the incident (and reported him missing), Travis was missing for five days—and then returned. The movie *Fire in the Sky* chronicles his ordeal, and Travis is still appearing at various events and telling his story today.

While Travis's case was widely known, the case that gained even more publicity was the 1985 case of Whitley Strieber. Whitley, a prolific author of many books, includes his most famous and profound work, *Communion*. It was not just the intriguing material in the book that made such an impact, but also the cover which depicted a typical gray alien, which has since become the most widely reported species seen today. *Communion* is one of the most important and historic chronicles of the subject. Strieber has written many books on the subject of alien abduction, and has also appeared on numerous radio and television shows around the world. Even today, he still has experiences and appears on various programs. *Communion* was made into a major motion picture which has helped us gain insight into these "grays" and the purpose of their visitations.

Fast forward to the late 1980's and early 1990's: This was the era when I became involved with alien implant research. MUFON had just begun to accept the alien abduction phenomena as a reality. The late Dr. John Mack, a tenured professor at Harvard University, came out

with his book, *Abduction,* in 1994. This text, in my opinion, led to the acceptance of alien abduction as a genuine facet of the UFO experience.

My historical contribution (as it turned out to be) was based on an attempt to expose a possible fraud, followed by the idea of "what if?," and then by taking a leap of faith, before it all fell together. Did a certain number of abduction victims have objects implanted into their bodies by alien beings? Was this a local phenomenon, or was it regional or worldwide? My instinctive nature led me to believe the abduction phenomena was a reality, but I could not wrap my mind around the idea that some individuals actually had these REAL-LIFE 3-DIMENSIONAL OBJECTS in their bodies. In actuality, it was my *disbelief* that propelled me forward to begin these investigations.

In 1995 I was writing articles for the official periodical of the Ventura-Santa Barbara MUFON group, *The Vortex.* Some of my investigative reporting included covering presentations at some of the official conferences on the UFO subject. I did as I was instructed, reporting my own flavored version and thoughts as to what I was hearing and seeing. Some of the presenters may have been displeased with my opinions, but nevertheless, they were my own and I was not dissuaded by negative comments on the matter.

During one of these events I was approached by someone with a set of x-rays of the feet: He made claims that the two apparent objects residing in the left great toe were ALIEN IMPLANTS. It was all I could do to contain myself from laughing out loud. I muttered, "Oh, wow!" and walked away thinking I had just come across another nut case. Within a few short minutes, the State Section Director of our MUFON group approached me and asked if I had seen the x-rays of the alien implants. I gazed at her in amazement, thinking she was much more naïve than I thought. I politely told her that I had viewed the x-rays and thought that they were very good x-ray views of the feet.

She inquired, "But what about the implants? Didn't you see the implants in the big toe of the left foot?"

I explained to her that, in my opinion, as a practicing foot doctor, they appeared to be two metallic objects that were probably leftovers from a previous foot surgery. I also explained that many times we use wire, pins, screws and other metallic hardware devices to stabilize bones when we operate. With that, she begged me to go back and take another look at the x-rays and talk to the individual who showed them to us. In an act of kindness, and with respect for my friend, I relented and accompanied her back to the table where this gentleman appeared again with x-rays in hand.

The initial conversation was brief and to the point. I inquired as to when the patient had undergone a previous surgery. He denied that the woman to whom the x-rays belonged had ever had an operation. I asked if he had a copy of her medical records and with one swift move his hand dove into a large suitcase-like bag and pulled forth a large folder containing a voluminous amount of papers.

"Here they are, if you would like to look at them," he challenged.

I asked if he would trust me to take them to my room and go through them. He agreed and I placed them in a plastic bag I had acquired at the conference. That night, I went through the records and noted that there were no entries indicating a previous foot surgery. I returned the medical records to the gentleman the following morning, suggesting that if he felt the objects on the x-rays were unusual, then he should see that they be removed and analyzed. He revealed that his client had no medical insurance and could not afford to pay for the surgery or the analysis. At that moment, I contemplated my next move. This might be an opportunity to prove this subject of alien implants was all a bunch of nonsense and have some fun at the same time. I asked the gentleman where this "implantee" lived and learned that she hailed from the great state of Texas. My offer was simple and genuine, "Tell ya what, you get her to California and I will do the surgery free of charge." Though he was surprised to learn that my offer was serious, we exchanged contact information and, after advising the section director of what had taken place, I left the conference.

Within a week of arriving home I received a phone call from the implant researcher asking me if I was able to do another case as well. Shocked, I was caught off guard, to say the least. I asked him where the object in question was located and learned that it was in the back of a fellow's left hand. Without giving it much thought I agreed. I do not perform surgery on the hand, so I had to enlist the services of friend who was a general surgeon. He, also, was intrigued and thought he would be able to share in the laughter with me when we extracted one of the many "terrestrial" foreign bodies we had both removed so many times over the years.

In August of 1995 these first two surgeries were performed. At the time I had no way of knowing this event would change my life forever, or that the hand of fate would turn the tables on who was to get the last laugh—for it was certainly not my colleague or myself!

A large, gray T-shaped object was extracted from one side of the great toes of the female patient, and then an object about the size of a small cantaloupe seed, but with a dark gray coating, was extracted from the opposite side of the same toe. The male patient yielded a tiny seed-like object from the back of his left hand exactly matching the one removed from the toe of the female.

Both surgeries also brought some very high strangeness. Neither of the surgical patients had any visible scar where the object would have penetrated the body. In addition, there was absolutely no disruption of the subcutaneous tissues, especially where the largest T-shaped object resided. This implant measured approximately one-half centimeter in each direction. This was large in comparison to the size of the toe in which it was placed.

It should be noted that the first surgery was done under a local anesthesia, as well as a hypnotic state, for relaxation purposes, performed by a professional hypnotist at the patient's request. During the surgery, at the point the foreign object was touched, the patient, in severe discomfort, rebelled and acted as if there had been no anesthesia administered at all. Both the hypnotic state and the local anesthesia had to be performed again. Those events, coupled with the fact that all three objects were coated with a biological substance that could not be cut through with a surgical scalpel, made the case not only "un-laughable," but very serious.

When the procedures were finished, I pondered about what the next step should be, and finally decided I would treat these cases as any other routine surgical extraction of a foreign body – I sent the surrounding soft tissues for pathological analysis and had no expectations of getting a report back with anything unusual. This was another error in my thinking. The reports indicated there was no visible sign of any inflammatory or rejection reaction by the surrounding tissues to the metallic foreign body. *This was not possible!* I thought I must not have kept up with some of the latest findings in pathology over the past thirty years of my practice. Immediately I went on the Internet and gained access to some of the world's most prestigious medical libraries such as Stanford and Harvard medical facilities. I learned the searches I had asked for were not free of charge, but actually quite costly. I had taken the quest this far, so I decided to spend the money and see it through, sifting through reams of the latest pathological materials of rejection reactions with metallic foreign bodies. I found new hormone responses and tissue activating triggers, but "lo and behold," there were no actual new findings from what I had learned in the past. This intensified the mystery.

It was my decision to have another pathology laboratory look at the specimens. This was done and the results garnered from the second lab agreed with the first. There was another factor that was a bit of a mystery: all the tissues surrounding all three specimens showed an intense amount of nerve tissue. I decided to home in on this finding and found that the nerves were a category of proprioceptors. These are usually found in the fingertips and the balls of the feet and are for the purpose of sensing pressure, fine touch, temperature etc. *What in the world would they be doing in soft tissue surrounding foreign objects removed from the side of a toe and the back of a hand?* I found myself being sucked deeper and deeper into the depths of a mysterious force for which there was seemingly no return. I had to know more!

There had been no attempt so far to look at the metallic portion of the specimens. I knew absolutely nothing of the study of metallurgy. All I could rely on was a very basic recollection of what I had taken in my college chemistry and physics classes—and that wasn't much. The next step was to seek advice of the professionals, so I secluded myself in my office and spent hours on the telephone trying to contact a person who could do some analysis. As it turned out, there were a few good-hearted souls who offered their services, but they warned me that they could not give me written reports because then I would have to bear the costs of "coming through the front door"—so to speak. These costs were far more than I had budgeted for a JOKE. I decided to give the back-door research a try and sent one specimen off to one of my friends at a local lab.

I waited for weeks and never received a response. My phone call to the individual caused another source of distress as he told me that he could only do so much in a day and I would have to be on the bottom of the list. I was counseled just to "be patient" and wait for his results. I thanked him and resolved myself to a long wait. I was not disappointed.

After nearly a month and a half I received a phone call from my friend who, in the middle of his analysis, wanted to know if I was playing a joke on him. I assured him that this was no joke and asked him what his findings were. He told me he likened some of the metallic properties to iron meteorites. I was flabbergasted, to say the least. I thanked him for his endeavors and tried to convince him to write me a report on his letterhead. He refused and told me he could not do this because the company he was working for would not allow him to have his own letterhead used for findings produced on the company premises. With that, our discussion was concluded and so was the matter of further metallurgical research.

As much as I tried to keep the word from getting out about the surgeries I had just performed, the more questions were asked from outside sources and particularly from the local MUFON group. I endeavored to keep my name a secret for a variety of reasons—especially because I did not want my professional reputation in the medical community to be questioned. Despite all my efforts, I was contacted by a lady who claimed to have had alien abduction experiences and told me that she thought she might have an implant on the back of her left neck and shoulder area. I questioned her in-depth on the telephone and finally agreed to a face-to-face meeting in my office. I listened to her stories pertaining to her strange dream-like sequences of her interactions with non-human beings, gray in color, less than four feet tall, with large heads and very large, black, upturned eyes. She was certain these were not just dreams as one morning she awoke with small spots of blood on her bed sheets and a very sore spot on her left shoulder. Looking in the mirror, she was able to see a small, raised, dime-sized red lesion in that region that was painful to the touch. She had not had any irritation in that area prior to her dream that night. She rationalized that she must have been bitten by some kind of insect in the night, but over the next two-week period, she noticed that the lesion had shrunk to the size of a pinhead—yet was still very painful to the touch.

During our visit I took detailed notes and told her to contact me in about a week. I also advised her that she should have an x-ray taken to see if there was a possible foreign body below the area of the visible skin lesion. Before she left the office I supplied her with a thorough questionnaire that I had compiled from other alien abduction researchers and asked that she answer and return them in the mail as they would help me to evaluate the case. She did as instructed and I received the package back in about one week. She scored in the high probability profile range for positive alien abduction.

During the next week the patient had an x-ray taken of the area and then brought the x-rays to me for my evaluation. I was surprised to see there was a round shadow below the site of the skin lesion indicating a non-metallic foreign body. At that point I called a colleague who was a dermatologist and a pathologist. I arranged for him to see the patient, and after the consultation he called me and explained that his diagnosis was a lesion called a calcifying epithelioma, based on his past experience. At that point, I did not inform him that the x-ray demonstrated a rounded radio opaque shadow deep to the skin lesion, but rather asked if he would consider surgically removing the lesion. He agreed and volunteered to perform the pathological analysis. I concurred, with the exception that if there was any hard foreign material removed that it would be turned over to me for further analysis. With those details out of the way, the surgery was performed a short time afterwards. As it turned out, a small grayish-white ball was removed along with the skin lesion. My colleague thought it was probably just a calcium deposit. I asked him how he arrived at that opinion without any analysis being performed and he told me that there was nothing else it could be. I had not decided what to do with the specimen yet, but knew that I must secure metallurgical analysis before making any further statements about the event. Also, I was prompted to do some in-depth research into the subject of calcifying epitheliomas and was amazed to find that the course of events leading to the formation of these lesions was exactly opposite to what the patient had described. Later, after metallurgical analysis was performed, we determined there was no calcium in this object whatsoever.

I did not openly speak about this case to many people outside the immediate circle of my friends or colleagues at MUFON. Most individuals I did confide in wanted me to push the research to a higher level with more cases and more biological research. Of course, all of this sounded very good, but was just too costly for my pocketbook and, without outside finances, I could not support the type of research that was needed.

During these discussions, contacts from outside sources began to increase.

My activities were now quite an interest to the procurers of news stories for the public. I made a serious effort to keep my real name out of the news, calling myself *"Dr. X"* and many other false monikers. Some of the publicity got out to the radio networks and I began to get calls from all over the globe from radio stations wanting to know if I was the one who had removed *alien implants from human beings.* Throughout this time I continued to be contacted by individuals who alleged they had been involved in the alien abduction phenomena. One lady called me from New York state and told me that she had read about a case that Budd Hopkins had been working on in which an abduction victim was floated out of a window in her apartment building up into a waiting "flying saucer" with two non-human beings, one on each side of the abductee. They were described as the typical gray beings with the big black eyes. She then revealed that she, too, had been abducted from a campsite in upper New York state by these same-looking beings. She described being on the ship which had numerous rounded corridors with rooms on either side. When she peered into these other rooms she could see other humans, naked and lying on tables while aliens performed procedures on them. She told me that she had no fear because she felt she had given them permission at a time when her soul was not yet born into her human form. She went on to say that they took her into a room and commanded her telepathically to remove her clothing and sit upon a table that had one pedestal in the middle for support. She did as she was told and was informed that the procedures they were going to perform would not hurt her and soon she would be returned home. Whenever she would begin to feel fear or pain, one taller-looking gray being would come close to her, put one of his big black eyes next to her eye, and suddenly all the bad feelings would be gone and there would be no more discomfort or pain.

The procedures she described were medical in nature. They did a vaginal examination with a long tube connected to an apparatus on the ceiling. The device had an end, which could be changed, and the new end-piece had a light on it. She went on to explain that the device was then placed over her belly button and slowly pushed in. She felt no pain and there was no bleeding. She added that they seemed to be looking at a television-type screen while moving this object around in her abdomen. She felt suction and a slight tugging and then watched as a small amount of fluid flowed into an attached container. When that procedure was finished, they examined her eyes and her tongue. She tried to ask the beings questions telepathically, but they did not answer except to tell her that no harm would come to her and that she had been there on previous occasions throughout her life.

I asked questions pertaining to the ship's lighting and structure, but her answers were vague. She responded that the lighting seemed to come from nowhere and was bright, yet subdued; and the walls were as shiny as metal, yet had an organic appearance. The temperature of the craft was cold, but not uncomfortable.

She described her breathing as being labored, and the air stale and oppressive. Her captors were somewhat mechanical, lacking in emotion, seemingly operating as a robotic unit. Occasionally she would hear a cry or scream coming from one of the other rooms; still, whatever method they had used to calm her took over her senses and while there, she felt no sympathy for the other humans.

Once they had finished with her examination, she was asked to get dressed and was then escorted out of the procedure room and into another area where other humans were sitting on bench-like protrusions from the wall. All sat in silence, waiting their turn to be taken, three or four at a time, through another doorway and into a room which she could not see from where she was sitting.

She waited her turn and was taken by two Grays through the passageway and into another large room which looked like a huge aircraft hangar. From that point her memory completely disappeared until she awoke in her bed at home. Upon arising, she felt sick with muscle aches and pains as if she had a case of the flu. Her primary need was for fluids and she ran into the bathroom and began gulping down numerous glasses of water. She also stated there were some red bumps and a tiny area of bleeding at her belly button site. The days following this event were filled with periods of depression, abdominal cramping, and irregular menstrual periods that lasted for several months. Since she was not married and had few close friends, she had no one to whom she could relate what had happened. This added to her depression and bad moods. Fearing she might lose her job if she lost control or shared her story, she did not bring these moods to her employment.

I was the first to whom she had confided because she thought that I could be of help to her. She thought she might have had an implant placed in her abdomen. The only way to determine if that were true would be to take an x-ray or perform a CAT scan of her abdomen. She was willing to have this done, so I proceeded to order the x-rays. They were taken the following week, but the report from the radiologist was negative for any foreign body. She seemed relieved with the result. Still, I mentioned the possibility of trying regressive hypnosis to learn more about the particular event she had relayed to me—or events that may have taken place earlier in her life. Our abductee followed through with my suggestion and the case became one that typically represents the alien abduction phenomena.

Another most interesting and typical case was that of a twelve-year-old girl whose family was having abduction experiences on a regular basis. The parents were stressed to their limit and feared for the safety of their daughter. I informed them that my field of study was that of material objects found in the bodies of abduction victims, but they had specifically sought me out because of my scientific approach to the entire subject rather than to find implants. I decided to at least listen to their stories out of academic interest on my part, and the compelling need on theirs. I was told that the phenomena had been going on for many years and that the parents on both sides of the family were also involved.

Their episodes began about three months prior to the birth of their daughter. The mother was taken aboard a UFO craft and some kind of procedure was performed on her abdominal area. She was in pain for several days, but when she was examined by her medical doctor, he could find nothing wrong. Otherwise, the pregnancy was normal.

Their description of the abduction scenarios was typical and also involved the gray aliens. Both the mother and father told me the Grays were not acting alone; the parents had also seen beings that look exactly like us with blonde hair and blue eyes. The mother described the male entity as "absolutely handsome"—to the point that it was difficult to believe that he and those like him were not human; they communicated with her telepathically, informing her that she and her family were not in any danger and that they were part of a worldwide experiment to alter the genetics of the human species. According to this "handsome alien," the daughter was going to be one of the NEW HUMANS. I had not heard this from anyone previously and thought it might offer some reason for the entire abduction phenomena.

In my discussions with the father, he told me that in his experiences he was connected to an apparatus that applied suction to his penis, then he was given visually-stimulating images of various types of human females in sexual activity. He was unable to control himself and felt as though he had ejaculated. No explanation was given to him for the procedure, yet the gray beings assured him that they were not going to hurt him. In addition, he was also "told" that he was participating in an experiment for the greater good of mankind.

The parents, rather than react negatively, accepted what they were told. Their attitude toward their abductors was not hostile, and they had learned to live with the phenomena—until their daughter was taken. They were greatly concerned when their daughter was aboard the ship because they were not in the procedure room with her and were unable to see what was happening to her and when she returned, she had little memory of the events.

The daughter differed from "typical" children in many ways. Her mental capacity was much greater than her school friends—to the extent that she often became embarrassed by her innate understanding of the world. She could calculate mathematical problems in her head at a very rapid rate and she had a great affinity for animals which, by the way, had no fear of her whatsoever. The parents spoke of a time when the daughter picked up a tiny, injured baby bird that had fallen out of a nest and was barely breathing. Within minutes the bird was re-energized and perky. The girl returned the baby to the nest without any objection from the mother bird.

I decided to interview the girl separately from her parents and obtained their permission to do so. Gracie was twelve years old with blonde hair and deep blue eyes and had a gentle disposition. She was slight of build and of normal height for her age. I explained to her that I was a researcher in the area of abduction phenomena and would appreciate it very much if she would talk about her experiences with me. She agreed to do her best to answer my questions. In general she described the following: Some time during the night she would awaken from a sound sleep in her room and would see a bright light coming through the window. During this time she felt an inner peace, but was wide awake and had no desire to sleep. Next, a ball of silver-white light would come floating over her bed, remain suspended there for a time, and would "smile" at her. This smiling orb would take her on numerous adventures where she could see the stars in the sky as she flew over the houses below. She could see the earth from afar, describing it as a beautiful "big, round, bluish-white ball just suspended in the blackness of the heavens."

The girl was taken to a place where there were many good friends who would play games with her—games that were a lot different than the ones she would play here. She tried to teach these "space friends" how to play cards, but they could not understand why she thought that was fun. Instead, she was taught different sorts of games. One involved a silver spoon-like object: the objective was to hold the "spoon" mid-air and toss it back and forth like a ball—but only using her mind, no hands allowed! Her new friends found this pastime great fun; however, when Gracie asked for a baseball and tried to gently toss it to one of the boys, he just ducked, letting the ball pass right by him without any attempt to catch it. Another task was to create a pyramid in mid-air, using their minds to add blocks onto the object until it got so heavy that it fell to the floor. When the blocks fell, they all laughed. I asked if there were any regular children there. She responded by asking, "Oh, do you mean from Earth?"

I was a bit shocked, as I had not anticipated this direct response. She told me that there were other Earth children there, but she couldn't tell which ones because they all had different appearances. She quickly added that here on Earth there are a lot of children that do not look the same either, having differences such as hair color, head shape, slant of the eyes, and so on. I acknowledged that fact and continued with our session.

I asked if there was an older person present who was responsible for the games the children were playing. She explained that at each session there were different people "in charge" and they were of different appearances. One of them, whom she described as mean in character, looked like a very large lizard with strange-colored eyes with pupils going across the eye. She thought he was very gruff, having little patience with kids, but he caused them no harm. She also told me that he had a very bad case of body odor, but she was afraid to mention that to the other kids.

Each of the sessions appeared to be some sort of lesson for both the Earth children as well as for the others. She did not know where the other children came from since open conversation was not permitted among them.

I was very interested in what she had told me previously, referencing how her abduction scenarios began by seeing a bright white light and gleaming bright silver ball hovering over the bed in her room.

"Could you draw what you saw in your bedroom?" I inquired.

Though she said that she was not very good at drawing, she said she'd be happy to try.

Her parents supplied a large pad of paper and some sharp pencils. I asked Gracie to begin the drawing with what she first saw coming into her room. I let her know that I would be asking questions from time to time as she drew to help me understand the image she was making.

She began her drawing showing that she was in bed, with the window in the far wall of her room. She drew pencil streaks coming in the window and told me that those were the beginning of the bright light. She then drew a basketball-sized object with pencil strokes emanating from all sides. I asked her if that was the silver ball she was telling me about previously and she nodded, yes. At that point she stopped drawing. I asked her what was going on now. Her response was that she could not remember what happened directly after the ball was over her. So I switched gears, asking her if she could sketch what was holding up the ball. She began to extend lines below the ball coming from each side. She told me these were the little legs it was standing on. I asked her if she could complete the drawing of the legs and she did exactly that by extending the lines downward and filling in the central section of each leg with some anatomical details including knee joints, and finally ankles and feet. I asked her if the ball also had arms. She immediately began penciling two skinny little arms extending from each side of the ball. After a couple of more questions about the details of the arms, she had added much more anatomical detail, including two hands with thin, elongated fingers. There were only four fingers and I asked her if she had made a mistake. She did not respond well to my question and seemed to think that I doubted the exactness of her memories. I calmed her fears by telling her she was a wonderful artist and I really appreciated the time she was giving me to draw this.

Next, she drew the final part of the drawing. I asked her if the ball had a head and neck and she began to draw from the top of the ball a very thin neck and on the top of this neck was a very large head. I did not ask her any further questions, but instead let her continue the drawing without comment. Soon she began to fill in the features of the face. I stood in amazement as she shaped a very tiny vestigial-type nose, a rudimentary mouth, no ears, and large black, haunting upturned eyes. I found myself staring into the face of one of the famous gray beings.

I thanked Gracie for her wonderful drawing and asked if it would be alright to show it to her parents. She was excited about showing them her picture and ran through the open door of the room, calling for them to "Come see!" what she had drawn.

I had no magic answer for the parents. The only thing I could offer this family was my past experience with other abductees whom I had interviewed, as well as the knowledge I had gleaned from the books of the great pioneers in this area such as Budd Hopkins, David Jacobs, John Mack and Yvonne Smith. I assured them that most cases of this type did not seem to bring irreparable harm to the abductees as long as they functioned normally in their daily living. I also advised them that if they wanted to seek further detail of these events that had transpired in their lives, then they should seek a qualified hypnotherapist for a regression, and if not, to just leave things as they were. I also reassured them that "they were not alone," but were part of some worldwide plan being carried out by an advanced civilization for unknown purposes.

By this time it was clear—these contacts and experiences had convinced me without question that we have hidden visitors. And so began my journey into the unknown...

CHAPTER 2
ENLIGHTENMENT

The word of what I was doing continued to seep slowly into communication channels I could not have predicted. Although I had made a concerted effort to keep my true identity unknown, a percentage of the public interested in the UFO subject apparently had a way of finding out how to contact me. As I mentioned earlier, I was not seeking this kind of publicity and thought that it might damage my career as a practicing member of the healing arts. Perhaps it was my active association with organizations such as *MUFON (Mutual UFO Network)* that served to garner the public's interest in the subject of alien implants.

Regardless, in a short time I had received many communications from individuals who claimed to have objects in their bodies and who were certain that they had been victims of alien abductions. I realized it was going to be necessary to put together a research team that would be qualified to handle the cases of those who were contacting me. It seemed that the serious nature of this entire implant subject should be treated as a true scientific study, and in order to accomplish this, a team of experts should be involved. When this all began, according to the available literature, only a small number of abduction cases had been discussed. These were primarily investigated and researched by Dr. John Mack, a noted tenured professor of Psychology at Harvard University; Budd Hopkins, a hypnotherapist in New York City; Dr. David Jacobs, a professor of Psychology at Temple University; John Carpenter, a social worker and clinical therapist based in Springfield, Missouri; Leo Sprinkle, an American psychologist who studied at the University of Colorado and earned his Ph.D. at the University of Missouri; and Yvonne Smith, a West Coast hypnotherapist who trained with Budd Hopkins. I telephoned Budd Hopkins and explained who I was and what I had accomplished to date with my research. He was very cordial and seemed genuinely interested in the subject of alien implants. He explained that his research had uncovered the possibility of objects placed into the bodies of abduction victims by non-terrestrials, but he had not had the time to scientifically delve deeply into the subject. I discussed the need for expertise in the matter of abduction pertaining to the specific cases of implantation and he advised me to contact Yvonne Smith. Strange as this subject appears to be, Yvonne Smith turned out to be a close friend of the State Section Director for Ventura-Santa Barbara MUFON, Alice Leavy. I called Alice and explained my need for Yvonne's help. She supplied me with her contact information and I wasted no time in contacting Yvonne. A meeting was set for us to discuss the matter in detail and I considered this to be a good start for securing a team of experts. My next efforts were to secure the medical experts needed for their surgical expertise. This was done on a case by case basis and over future months. One final member of the team had to be secured, and that was a psychologist with a Ph.D. open enough to be involved. I knew somewhere down the line, the subject pertaining to the psychological condition of the surgical clients was going to be paramount. After discussing this with Alice Leavy and a few members of the MUFON group, I came upon a lady who had recently acquired her Ph.D. in Psychology. The fact that impressed me the most was she had a very unusual topic for her Ph.D. thesis: Alien-Human Hybrids. I took little time in acquiring her contact information. Her name was Christianne Quiros, Ph.D. Our first contact was by telephone and a face-to-face meeting was set for the following week. I became increasingly excited as the scientific investigative team was actually becoming a reality.

My reception by Yvonne and Christianne was heart-warming. We sat and discussed the future proposed enterprise, realizing it could open up an entirely new pathway of research into the alien abduction phenomena—perhaps one which would supply the "holy grail" of the entire UFO subject! And what might that be? Our search would be for the scientific, physical evidence *proving* that human beings were interacting on a physical level with non-terrestrial intelligences.

With the research team now in place, I was able to proceed with the second set of surgeries which consisted of three clients chosen from a list of the individuals who had contacted me. They had all passed the preliminary tests qualifying them as potential surgical candidates. The three most important of these were the following:

1. Past memory of human-alien contact without hypnotic regression.

2. X-rays or CAT scans demonstrating a visible foreign object.

3. A report by a certified radiologist confirming the findings of a foreign body.

Furthermore, there were other major requirements such as a fully completed battery of written questions which we sent to the surgical candidate—their return was mandatory before any surgery could be performed. We also required them to be in reasonably good health, both physically and psychologically. And they would need to be willing to sign a release which stated that any object recovered during the surgical procedure would become the property of A&S Research, our non-profit entity. Above all, the prospective candidates must want the surgical procedure to be performed, understanding that there would be no guarantees forthcoming as to after-effects, beneficial or otherwise. Our protocol was and is that no surgical candidate would ever be coerced into the removal of any suspected implants. Any knowledge we obtain from the procedure and attendant research would be shared with the abductee—and ultimately the world.

In addition to all the above considerations, we also informed our clients (the possible abductees) that no charges would ever be made to them for any aspect of the surgery or laboratory analysis. This, however, did not include the cost of transportation from their home to the surgical site. In most cases we, however, would cover the cost of their stay, including local transportation and meals.

It was also explained to the surgical candidates that there is a rather rigid schedule for them to follow once they arrive because of coordinated events which must take place. Each surgical candidate must proceed through the following steps:

1. Initial history and preoperative physical examination by the surgeon

2. A Phase One psychological examination

3. Team of experts' examination of surgical area covering various scientific disciplines:

    a.   To determine any emission of electromagnetic fields and/or radio wave

    b.   To expose site to three frequencies of ultraviolet light

    4.   An immediate pre-operative x-ray taken and read by one of our radiologists

    5.   CAT scans as needed

    6.   Routine medical laboratory exams including, but not limited to hormone levels, specific hormone precursors, or antibody levels

The three new candidates for surgery were two females in their early forties and one male in his early fifties. All three had some memory relating to alien abduction. The male had only a few memories and stated he thought he was in communication with extraterrestrials as he heard audible voices. His psychological profile ranked in the norm. He told us he had a few dreams in which he described the typical gray alien beings, but most of his dream sequences were vague and without order. He also scored a 35 on our Possible Alien Abduction Profile which indicated he was on the low probable end of the scale. The male client's object was visible on x-ray examination and was located in the lower left jaw area; the women both had objects in the left leg—almost in the exact same place. Also, above the object was a visible scoop mark which has been accepted by researchers in this field to be only related to alien abductees. The mark was a small area where a portion of the skin appeared to have been scooped out with some sort of instrument; directly beneath this mark, a rounded BB-shaped object could be seen on the radiograph. My research into the origins of this mark was interesting: no description of this type of skin lesion was found in any of the texts on dermatology or pathology.

Both female candidates had specific memories related to the abduction phenomena. One of them, whom we will refer to as Anne, had an in-depth recollection of a typical abduction experience where she saw a bright light coming through the bedroom window of her mobile home in the early morning hours. Though she tried to move in her bed, she was paralyzed and could only turn her eyes. Two alien grays floated her out of bed and through the bedroom window. She couldn't see where she was going and was terrified! She then found herself aboard a metallic craft with other human beings. They were all naked and seemed to be in some sort of waiting area. Next, she recalls lying on a table which felt neither warm nor cold, but was solid and metallic. She described what appeared to be a medical-type apparatus surrounding her, and at that point, she felt her arms and hands restrained. Certain procedures were performed which were painful, yet when the pain became unbearable, one of the taller gray beings would come over and look into her eyes, producing a calming effect which would cause the discomfort to disappear. The next thing she remembers is being back in her bed and feeling ill with flu-like symptoms, including aching muscles, extreme thirst, dehydration, and headache. After arising in the morning and while showering, Anne noticed the mark on her left leg. There was no pain and she stated it was not there the night before her strange dream. There was no blood noticed on the bed sheets. An added point of interest in this particular case involves an account by the client's neighbor. The woman next door saw a very strange-looking round craft hovering over our client's home on the very night that she was abducted. The "saucer" was creating a "buzzing" sound that interfered with her television and phone reception. The craft emitted a bluish-white light which was aimed at the window of our client's home. When the neighbor was interviewed later in the investigation of this case, the account was proved factual in every aspect.

The other female who was our third candidate had a positive abduction history and also a scoop mark on her left leg, almost in the exact same spot as in Anne's case. We will refer to this lady as Doris: she lived in the San Fernando Valley of Southern California and some years prior had a conscious memory of seeing a UFO hovering over her home. This incident was also witnessed by her mother. Doris related another typical abduction scenario. She was convinced that these abduction-type events could have been occurring since childhood. Several years following a particular event, Doris was attending a UFO Conference in San Diego, California. She was involved in a mass abduction episode soon to be detailed in Yvonne Smith's new book, *CORONADO ISLAND INTRUSION: The President, the Secret Service, and UFO Abductions*. During this case Doris found herself terrified on a night in which several abductions were taking place in the same hotel.

U.S. President Bill Clinton was staying in a private residence near the abduction event. Throughout the night the Presidential Secret Service was active at the general abduction site and all during the next day's events.

The three surgeries were all performed under local anesthesia. There were no noted pre-op or post-operative medical problems. The patients tolerated the surgeries well and all were in satisfactory condition following the procedures. The object removed from the male client was a small triangular piece of metal covered with a biological coating seen previously in our other extractions. The metal was shiny and the surface smooth. The surrounding soft tissues were removed for pathological analysis. With examination under ultraviolet light, the object presented a fluorescent greenish color. The surgical specimen was then placed into a container containing the blood serum of the patient and would remain there until the laboratory performing the first biological test opened it.

The second surgery was performed on Anne. It was decided prior to the surgery to excise the entire scoop mark from Annie's leg and submit it for pathological analysis. This might give us a further clue as to the mechanism used to produce the mark. In addition, a deep incision was made below the abnormal skin lesion. A small grayish-white ball was found and isolated from the surrounding tissue. It looked like a calcium deposit to the naked eye, but without in-depth metallurgical analysis, it was impossible to tell its exact composition. The object also displayed fluorescence of a faded green color under the ultraviolet light. One of the interesting factors at that moment was the obvious lack of a biological coating, as seen with the other metallic objects. We instantly knew we had something quite different than the others. Annie's object was then placed into the serum container and sealed. Anne tolerated the surgery well and was in the recovery room with the first surgical patient. She had no untoward affects from the surgical procedure and was assured of a typical and routine recovery. One more surgery to go—and the mystery of the moment swallowed up the surgery room.

The last surgery to be performed was on Doris. She was calm and extremely curious about the object we were set to remove. She tolerated the injection of the anesthetic well and was prepped and draped for the surgical procedure. It was also decided to excise the entire skin lesion so we could compare them both superficially and with pathological analysis. When the tissue was examined, we noticed how similar both skin lesions were to each other. The shine that was present remained with the excised piece of tissue. Many times when a typical superficial lesion is removed from the patient it begins to curl before it is submerged in the preservative solution. In this case there was no degradation or deformity noted. The next step, as with the other surgery, was to increase the depth of the incision and isolate the little round ball we had seen on the x-ray. This was done and we conveyed a small grayish-white, BB-sized ball to the waiting container. As with the other specimens, ultraviolet examination revealed a greenish fluorescence.

At this point it is important to consider the findings of the pathology reports. The male client had a report showing no inflammatory response, no evidence of rejection of the metal, large amounts of nerve tissue in the area, and no infiltration of abnormal cells. This was consistent with the reports of the first two surgeries with metallic foreign bodies. Both Doris and Anne had very similar reports. One of the most interesting of the findings was that of Solar Elastosis—meaning the skin had been exposed to a large amount of ultraviolet light without burning. The facts are that both of our female clients were housewives without any indication of exposure to large amounts of sunlight. Even if this is only partially correct and both surgical patients were not telling the truth, then it would be difficult to explain how only a few millimeters of skin could show signs of overexposure to ultraviolet light without the entire skin surface being affected.

All three cases were followed for several years after their surgery. Immediately following the surgery and for the next several weeks the male client stated he had no more events occurring and the voices had disappeared. Later contact with our male patient revealed the voices had returned, but he could not report any further contact with non-terrestrials. Both females were followed for several years. One reported no further events had occurred and her life was now peaceful and quiet; the other, however, reported continued contact, but no further implantation.

An extraordinary event that did occur was on the fourth month post-op, to the very day, when I received phone calls from both Doris and Anne. They independently told me they were having terrible pain in the surgical area and it had turned a bright red in color. They both denied any injury to the area and stated that no perceptible abduction event had occurred. They both denied related dream sequences of alien contact prior to the increase in pain and redness in the surgical areas and were in different geographical locations when this happened. I told them it was important for me to examine them as soon as possible. This was arranged in separate visits to my office. In both cases the redness had subsided by the time I examined them. There was almost no discoloration of the area. Only a small scar was present at the surgery site. There was no pain discernible upon superficial or deep palpation of the area. Examination with ultraviolet light did not produce any fluorescence.

My research into this field has produced many mysteries, which to this day are not explainable in scientific terms. I will endeavor to point them out further in this text.

CHAPTER 3

WHERE IS THE PHYSICAL EVIDENCE?

How many times on radio and television have we heard the question, Where is the physical evidence in the field of Ufology? Isn't this the number one question asked by the paid debunkers and skeptics? Isn't this also the question inserted into the public consciousness?

After many years of scientific research, I have reached the conclusion that there is *no lack of physical evidence at all*: actually the converse is the truth. There is so much physical evidence that it is impossible for the corresponding amount of analysis to be done without a large expenditure of funds. This defines the problem in the nutshell. The physical evidence is abundant, but there is a pronounced lack of funding to even begin cracking the surface of scientific analysis.

My personal endeavors have taken me deeply into the field of alien abduction and particularly that portion pertaining to physical objects contained within the bodies of alien abductees, popularly classified as *alien implants*. Never before have I discussed at length the hundreds of cases of physical evidence that do not pertain to implant research. In this chapter I will discuss some examples.

A SUPPORTIVE PHYSICAL EVIDENCE CASE

Over ten years ago I was contacted by an abductee in the Southern California area who related a story of alien contact and continuous visitations resulting in large amounts of physical evidence not related to alien implants. I did my best to explain that my specialty and scientific investigations were only related to alien implantation; however, he had acquired so much material and insisted that his experiences were so phenomenal that he finally convinced me to become involved. This was the beginning of a long and lasting relationship which opened my mind to other non-implant forms of physical evidence.

The gentleman, whom I will refer to as "Glen," was in his forties, married with two children, and was trained as an arson investigator for a fire department in a medium-sized Southern California community. He explained that some years prior he was awakened during the night and beheld non-terrestrial entities in his bedroom and in other parts of the house. He related to me a complete laundry list of abduction events, many of which were the typical abduction scenarios. He talked about seeing his bedroom filled with a bright whitish-blue light, having his wife who was sleeping next to him "switched off" and non-responsive during the abduction event, and being taken onto a craft either with a beam of light or actually escorted by two or more non-terrestrial beings. His experiences on the craft were varied and happened so frequently that the logging of these events became confusing.

In the beginning Glen tolerated and coped with these experiences satisfactorily, yet as time wore on, coupled with the demands of his occupation and his lack of sleep, problems began to arise at his place of employment. His immediate superiors were concerned for his welfare and safety. They recommended he have a consultation with a psychologist and supplied him with the name of someone in his area that the fire department had used previously in cases of Post-Traumatic Stress Disorder. It was obvious they thought this was the root of Glen's problem.

After several visits the psychologist suggested that Glen undergo a hypnotic session to help get at the deep-seated root of the problem. Glen, without question, agreed. The following week, Glen arrived at the psychologist's office not quite knowing what to expect as he had never before undergone hypnosis. The doctor explained the procedure, then began. Glen settled into a quiet hypnotic state. The next thing he recalls upon waking is the doctor telling him his session was most interesting, but he probably could not help him. Though Glen was startled by the remark, what he heard next unsettled him in a way that would affect him for the rest of his life. The psychologist recommended he see someone specializing in the alien abduction phenomena. Glen was in an instantaneous state of shock. He asked the doctor numerous questions, but all his answers were unsatisfying.

More psychologically upset than before, Glen confided in his wife, but she thought "that stuff" was a bunch of nonsense. Glen decided his life could no longer continue the way it had been going. He started researching UFO's and alien abduction on the Internet. The more information Glen found, the more answers he sought. He decided at that point that he would document his experience by collecting physical evidence.

Since his occupation as an arson investigator had trained him in the search for clues, he knew instinctively some of the directions he needed to go. His first act was to install a small video camera in a clock radio in his bedroom. The camera was motion activated.

Glen's experiences continued after installation of the camera, but did not capture any images. Some streaks that should not have been there were found on the video, but their cause could not be determined. During the time of the filming, Glen had an experience of being taken aboard a craft by a group of non-terrestrial beings not described in the conventional abduction literature. They were short in stature, muscular, with a large amount of body hair. Their heads were large and faces appeared beak-like. They communicated telepathically. These beings told Glen that the future of this planet was in jeopardy and turbulent times lay ahead for humanity.

He was shown three-dimensional scenes of devastation, including one of the east coast with New York laid to waste. The root of the catastrophe was not evident, but the entire region was devoid of population and the Statue of Liberty was lying on its side in the water. After Glen was shown this despairing scene, he was taken out into deep space, and then was returned to his home. These beings gave him no reason for his abduction, his trip to space, or the visuals they had shown him.

Glen continued to search for more physical evidence, poring through hours of video taken with his hidden camera. He later purchased a device called a frame splitter which tape-records video images at the rate of thirty frames per second. Each of these frames is composed of two visual fields. The new device he acquired would split all the footage into twice the number of visual fields. This is when his first breakthrough occurred. He spent hour upon hour reviewing these visual fields and then suddenly, there it was! About six frames showing an entity coming through his bedroom wall! This is one of the first pieces of physical evidence which Glen had supplied me. The entity was like nothing I had ever heard described previously. To say the least, it was ugly and menacing in appearance—at least by human standards.

The very first thing I did was to send these images to video analysts I trusted. At that point my 501(C) 3 did not have its own video analyst on our science board yet. In a few weeks I received the first reports. To my amazement, the consensus was that the images were genuine—*not hoaxed,* and indicative of a genuine entity none of us had ever seen before or heard described in abduction literature. I called my dear friend Budd Hopkins in New York and consulted with him on the phone. Budd was considered by many to be the world's leading authority on alien abduction. I had also sent him photos and he told me that in the vast amount of data he had collected over the years, there were no entities such as these ever described.

I have shown these images to audiences all over the world and have never had anyone tell me they are familiar. Another aspect of Glen's experiences involving these creatures is that most of the experimentation they have done on him occurs in his own bedroom. He told me they sometimes come in carrying equipment. This includes boxes, as well as other instruments and devices. He also received injections of materials which have left puncture marks on his skin. Many times he has awakened in the morning with symptoms of dehydration, thirst, aching muscles and other symptoms involving his digestive tract. He is told nothing by his abductors as to what they are doing or what they hope to accomplish. Glen describes their activities as mechanical and believes they consider him only as an experimental object. They show no emotional concern. In many instances, when these entities get ready to leave the confines of Glen's house, they reach down to a rectangular device attached to some type of a tool belt, press a button or switch and wait until the entire visible environment begins to oscillate in a vertical plane which appears similar to translucent vertical venetian blinds. Then they walk through this optical distortion and disappear.

Glen's other experiences have involved the stereotypical gray beings as well as reptilians and some who appear very human. On certain occasions Glen is not taken through a wall or ceiling but directly out of his front door and onto a waiting ship, landed nearby on his property. He called me about one instance when a ship landed on a gravel portion of his property and left oily indentations in the base gravel. I instructed him to gather up some of this material for later analysis. He complied and we still have the material, but unfortunately there have been no funds forthcoming to do the analysis.

In another related incident, Glen was awakened in the wee hours of the morning by two entities standing by his bedside. He described these as looking slightly like the gray beings, yet taller and with slightly different faces. The room was lit with a very bright light. He was instructed to get out of bed and accompany them. Glen tried to wake his sleeping wife so she could get a look at the beings, but this was to no avail. He did as they commanded and started to walk out of the bedroom. He knew from past experiences they would erase his memory and he would not be able to recall anything except bits and pieces of his experience. He knew he needed some kind of device to jog his memory when he returned so he could at least prove what happened to his wife since she was still in a state of denial. As he was passing through his living room, a thought came into his mind. He pretended to have a coughing spell and knelt down on the floor for a few seconds. During this time he removed his wedding ring and stuffed it into his wife's shoe which had been sitting on the floor. He quickly got up and accompanied his escorts.

The following morning, Glen was in the process of getting ready for work. He showered and was about to shave. Suddenly he noticed his wedding ring was not on his finger. He was in a state of panic as he could not imagine what happened to it. He was also afraid to mention it to his wife in fear she would blame him for the loss. He proceeded to get dressed and left for work without a mention of the missing ring to his wife. Glen received a telephone call later that morning from his wife. Her mood seemed strange on the phone. He asked her what the purpose of the call was as she was not in the habit of calling him at work. With that, she asked him if he knew the whereabouts of his wedding ring. He knew better than to try and make matters worse by lying so he blurted out the truth by telling her he had lost it somewhere during the previous night. He also asked her why she called him at work in reference to the ring. To his surprise she told him she found the ring in a very peculiar place, which was inside of her shoe she had left on the floor in the living room. Instantly Glen's mind was propelled back to the previous night's events and he explained everything to her. At that point she started to become a believer.

Glen called me following this incident and told me he had decided to move to another area in Southern California and he would like me to come over before he was completely out of the house and look for physical evidence that could be recorded. I arranged for a date to do this.

About this time, I was contacted by a Japanese television production company that was interested in the abduction scenario and was expressing interest in physical evidence. I had previously worked with this company and it was one of the major TV enterprises in Tokyo. They had filmed one of my surgeries and the results of the scientific analysis. This time they requested information on non-implant evidence. I explained Glen's case and they asked for permission to do some filming in his house. With that I called Glen and explained the situation. He gave his consent and a date was set for the shoot.

Glen resided in a one-story framed stucco home. It had a large yard and a circular driveway. Part of the driveway was paved with asphalt and the other part was composed of small pea gravel. There were a few sparse trees in the yard which were no more than six feet tall. The house had numerous windows all facing the street and front yard. The back yard was not large. Both front and back lawns appeared dried out, possibly due to the time of year or a need for water.

The interior of the home was typical for the homes in that area and age. Most of the furniture had already been moved. Basically, the only items left were one chair in the living room and a wheelchair in one of the back bedrooms. The floor plan allowed for entry into the living room; attached was a kitchen and dining area. At the other end of the living room was a door which led to a rather long hallway. Off that hallway was a series of bedrooms and one bathroom; plus another bedroom with additional bathroom elsewhere.

While the TV crew was setting up their equipment in one of the back bedrooms, I was out in the kitchen area with my detection instruments. The first items checked were the counter tops and the drawers. I used a Gauss meter, which would detect electromagnetic field strength up to 10 milliGauss (mG). The counter tops were made of formica and immediately I picked up electromagnetic radiation of 6 milliGauss. Next, I went to the kitchen drawers and each drawer was magnetic in varying strengths depending on the contents. The silverware in the drawers had a reading of 8 mG. Plastic salad serving spoons and forks showed a magnetic field of 4 milliGauss. On the bottom of one of the cupboards was a large rounded cardboard can containing latex interior paint. The cardboard container registered 6 milliGauss. The metallic hinges on two of the upper cupboards demonstrated a magnetic field of 6 mG. I then used a magnetic compass to learn more about the magnetism coming from the cupboard hinges, and to my surprise, I could not find a distinctive negative and positive pole. Both hinges seemed to be some type of a uni-pole. I then removed one of the hinges for further testing in the lab and it was verified that there was only one detectable pole. I have no scientific explanation for this, even in light of recent scientific breakthroughs.

An ultraviolet light was used to examine the interior walls. In the master bedroom there was a wall which showed an approximate one by two-foot area which contained some fluorescence in a yellow-pinkish hue. There were also numerous areas of fluorescence around and on the closet door. This was speckled and scattered on various parts of the door. I was told by Glen it was in this area where he saw the alien beings entering his room. "They always come through the closet wall," he remarked.

I asked him if this was the famous closet where he had found the "CLAW," and yes, this was the place! Glen then retold the story about how he collected this mysterious object. His quest for physical evidence became overwhelming. His mind was working overtime to think of new ways and devices to accomplish this purpose.

Glen concluded that if these entities come into our three-dimensional world by using some advanced technology, they must still conform to our three-dimensional environment. One of these is their physical response to our gravity. He concluded they must have mass and weight and therefore would leave footprints just as we humans do. To this end, he decided to use a towel the same color as his bedroom carpet and laminate several layers of thick tin foil to the underside of the towel. He then placed it on the floor of his closet next to the wall most used for entry by the alien beings. He made sure the towel was freshly laundered to ensure the footprints would be accurate. The first morning after the placement of the towel, he was disappointed as there were no prints. The second night of no results was even more disappointing, but he also realized he had not seen any beings entering his room those nights. By the third day he was even more discouraged as there were no visitations and no footprints. Each morning he diligently examined the towel. Suddenly after a fitful night of tortured sleep, he awoke with a headache and feelings of slight nausea. He stumbled into the bathroom and vigorously washed his face. He was extremely thirsty and drank several glasses of water in rapid succession. This was a work day so he decided to put his physical complaints on the back burner and get on with his preparations for work. He had almost forgotten about the towel. When the thought finally crossed his mind, he ran into the closet and was about to pick it up when he noticed it had been moved slightly from its original position. He stopped and turned on the closet light. His effort had been rewarded: there were two depressions on the towel. He carefully removed the towel and laid it and the tin foil package on the bed. "Yes!" he thought." My mission was successful." He had captured one complete and one partial footprint.

I clearly remember the day he called and told me the story on the phone. He was excited beyond belief and told me he was going to pour a plaster cast of the print and eventually bring it down and show me. He also told me of his plan to again place the towel so that more prints could be captured. He reset the towel, but did not receive the reward he had sought. He was only able to capture some scratches in the foil, which he concluded were made by an elongated toenail or claw. Glen continued the placement of the towel for the next couple of weeks. During this time he came to my office and brought the towel device he had constructed as well as the cast. The footprint appeared small and narrow. It seemed almost human but with some subtle differences. An impression of a human weight-bearing foot usually shows weight distribution from the heel area and along the bottom-outside area of the foot, which then crosses over into the toe area from the outside of the fifth toe joint to the great toe joint. The weight-bearing pattern of this foot demonstrated weight-bearing from the center of the heel, then forward directly to the ends of the 1st and 2nd metatarsal heads. There only appeared to be four toes and in general, the print appeared child-like.

Just before Glen was going to discontinue the experiment, his wife noticed a small dark-colored object caught in the thread weave of the towel. She was going to wash the towel as she had done each day at Glen's request. She called Glen's attention to the dark object and he took tweezers and removed it. It was dark in color, extremely hard, and measured about three-quarters of an inch in length. It was sharp and beak-like at one end, and the other end appeared blunt and squashed.

Again, Glen called me and reported his latest finding. I instructed him to put the object in a sealed container and to get it to me as soon as possible. Within a week he brought it to my office and I saw it for the first time. It looked exactly like some type of claw. It was my opinion the creature that had entered Glen's domain had possibly gotten it caught in the threads of the towel and had lost it in the process. I immediately started a literature search and found only a couple of instances where abductees or contactees had described a claw on fingers or toes. At that time I did not realize the extent of the research upon which I was about to embark. Almost everyone of a scientific background to whom I showed the "claw" agreed it indeed was exactly that—a *claw*. Each offered suggestions on what should be done next. One of them led me to a researcher and scientist at the San Diego Zoo. This Ph.D. was an expert in the field of Primatology. I spent almost one entire afternoon with this kind, knowledgeable and sincere lady. She pored through advanced voluminous texts and examined every physical feature of the suspected claw. She told me that certain species of primates did have a claw, which was called a "grooming claw," used for that purpose. She also told me she believed there may have been an internal structure within our object that could have been a poison gland. She was not able to tell me what creature or even what kind of a primate this had come from, but was kind enough to put her observations in writing.

I then contacted a friend who was one of the world's leading geneticists and explained the problem to him. I sent him photos of the object and he also thought it looked like some sort of claw. It was his opinion, the only way to determine what species to which it belongs would require DNA analysis. He explained to me he had a small fund set up and would be willing to discuss the matter with his colleagues to see if he could use that money to discover the source of the claw. I was overjoyed in hearing his plan and informed Glen what was about to happen. In addition, I needed Glen's permission to cut through the object so we might have fresh DNA sources. Glen did not hesitate and wholeheartedly gave his permission. Only a few days went by when I heard from the geneticist. The project was a go. I cut and sent the specimen to him via Federal Express. In a few weeks I received a phone call from Australia. It was my friend who told me he was having trouble with the DNA primers, as he could not find a match to the extracted DNA. He also told me he was running out of funds and it was going to cost more than he anticipated. In addition, it would be more complicated than he originally thought. At that point, I instinctively knew more help and more expertise was going to be needed.

I called my old friend at the National Institute for Discovery Science, Robert Bigelow's organization, who had helped me in the past. Colm A. Kelleher was a Ph.D. biochemist who had a good working knowledge of DNA Research. I told him about the claw and it caught his scientific interest. His base of operations was still Las Vegas, so one day I found myself, with claw in tow, on a plane to that city. Colm examined the object and thought it was of definite scientific interest. I put him in touch with my geneticist in Australia and at a later date this resulted in a face-to-face meeting at Colm's office in Las Vegas. During this meeting I sat and listened to these two scientific giants discuss the intricate minutia of the science of DNA. I felt as small as an ant since their discussion was far from my field of expertise. I felt honored just to be present. Their final discussion led to a plan. Some funding would be provided by NIDS and my Australian friend would carry out the further agreed research. Colm was to write an article for a peer reviewed scientific journal based on the DNA research performed on the claw. The research would conclude that the claw came from an unknown earth species. I informed Glen about the proposed plan and he agreed it sounded viable. I was excited, as this would be the first opportunity to get a paper published on the subject of UFO's in a scientific peer reviewed journal. It should be understood by the reader, however, that the plan was not to mention anything about UFO's or aliens in the first article, but only to get the gist of the DNA research in front of a scientific audience. It was our plan to then publish a full article in another journal illustrating all aspects of the case, which included the subjects of UFO's and aliens.

Several weeks went by and I received no communication from either Colm or the geneticist. One quiet afternoon, I was on an office break and was told there was a phone call from Australia. My secretary told me the party said it was extremely important. I took the call and it was my geneticist friend. He asked me if Colm had written the article yet and if he didn't, I was to tell him not to write it. He told me they had just developed some new DNA primers and it was concluded that the object in question was not a claw at all, but rather a *SLUG* (like a snail, but without a shell). He told me he would fax me the results of the latest DNA information and to please get in touch with Colm right away. I did as he requested. I reached Colm in time to stop the article from being printed. If this had happened, we would have looked like complete fools.

I was devastated and did not know what my next step was going to be. In a phone call to Colm, after he received the DNA results, he advised me to see a specialist in slugs and seek confirmation of the DNA findings. I located a Gastropodologist at the Museum of Natural History here in Los Angeles and made an appointment with him.

I brought the object we thought was a claw to the specialist in the field of slugs. He examined the object and showed me the anatomical minutia verifying it was indeed a slug. I asked him if he could tell me what kind of slug it was and he explained, in its present state, he could not. I showed him the print-out of the DNA analysis indicating a 97% chance it was a Sea Slug and a 93% chance it was from New Zealand. He explained he could not argue with the results. I also asked him how the animal became so hard as if it was almost petrified. He told me it was common and the slug just probably dried up over a long period of time. I could not agree with that opinion as it took great pressure to cut it with a scalpel blade.

In an attempt to solve the petrifaction mystery, I asked Glen to gather some local slugs and put them in a dish in the open sun and see what would happen to them. He did as I suggested and we found that petrifaction never took place, but rather they turned to powder, then dust and you could blow them away. The Gastropodologist was dead wrong on this issue.

Now the incident of the claw-slug was an even greater mystery. One would have to ask the following questions:

1. How did a sea slug travel from New Zealand to Glen's closet in Southern California?

2. How did the slug become petrified?

3. What is the relationship between the slug and Glen's ET experiences?

Our Science Advisory Board at A&S Research is composed of some of the world's greatest Ph.D. thinkers and they, too, are baffled. Is it possible the human race is interacting with advanced intelligences that have some sort of a sense of humor? We can't forget that Colm Keheller Ph.D, was involved with the research at the Skinwalker Ranch owned by Bob Bigelow. They evidently came to the same conclusion. They were being out-maneuvered and out-thought at every turn of their investigation. I have known abductees who have posed the same question to me. It seems, when asked, abductees have two belief systems. One is that these gray beings have no emotions at all, and the other is that they do have emotions, but don't understand human emotion. One abductee who was in his eighties told me that he tried to tell his gray captors a joke and they didn't get it, but in a telepathic conversation he caught one of them asking him why he was telling them a joke. To that he replied telepathically, "Aha! Now I got ya! If you didn't know what a joke was, then how did you know I was trying to tell you a joke?" The answer came quick and simple—his non-human friend replied simply, "Because you were laughing in your mind when you told it." So much for "one-upmanship" with a non-terrestrial.

The Japanese television company continued its filming in the rear portion of Glen's house. Glen was engaged with the production crew and his wife Helen and myself were left in the living room. Since the house was devoid of furnishings except for one chair, we sat on the floor facing each other. The door from the living room led into a long hall which had the various bedrooms and bathrooms jutting off at a variety of angles. I had presumed the TV crew was interested in keeping the area they were shooting in as quiet as possible, so I believed they had shut the door to the living room. Also, it is important to mention because of the age of the home, all the door hardware was on the old-fashioned side. It was necessary to have a good grip on a door knob before attempting to turn it and when turned, one discovered it had to go through a complete rotation to release the door mechanism that would allow the door to unlatch.

Our conversation was kept at a low volume so as not to disturb the filming activities. Helen and I were in the heart of a discussion involving the modern children of today and how different they were from folks my age. She was in agreement with my statements and was giving me a number of examples of what her own children were capable of saying and doing. Suddenly our attention turned toward the hallway door. Our conversation ceased and we sat and watched as the door knob started a full rotation, which would allow the door to open. We anticipated seeing the door fully open and a member of the film crew enter the living room. Nothing further happened. The door stood slightly ajar. I jumped to my feet and opened the door fully with Helen standing behind me. We peered down the the long hallway. All was quiet and still. We heard the conversation of the film crew in one of the back bedrooms. No one was in the hallway. My hair was standing straight up on the back of my neck. I turned and looked at Helen. We looked at each other in shock; our thoughts were clearly shared, almost transferred telepathically. Was there a ghost-like paranormal phenomena in that house? Did this have anything to do with Glen's abduction experiences? Little did I know, the answers to these questions would be directly forthcoming.

The television crew came into the living room and asked Helen to participate in the filming. I asked if I could occupy the back bedroom as I wanted to stay out of the filming procedure. Glen told me there would be no problem with my request, but the only seat in that room was in a wheelchair which he had not moved out yet. I assured him any chair was just fine and thought of myself catching a few quick winks as I was becoming tired from the day's events. It was now about 4:00 PM in the afternoon. I proceeded to the back bedroom, shutting the living room door behind me. Glen was correct. The only piece of furniture in the room was an older model wheelchair. It seemed sturdy enough. I turned around and slowly sank into the seat. After a couple of light bounces, I concluded the chair was safe to occupy. I had decided to leave the bedroom door open in case our phantom door opener decided to make a return performance in the hallway. Since the wheelchair appeared to have good functioning wheels, I decided to take a little ride around the room. Oh boy, I was actually having some fun and passing the time with some upper body exercise. I tried a few fancy maneuvers since I had watched television shows on handicapped basketball players who used wheelchairs. On one attempt, I accelerated too fast and almost slammed headlong into the distant wall. At that point I decided to just rest. One thing I had been wanting to do was to clean some of the papers out of my pocket-sized checkbook. This would be the time. I removed the checkbook from my shirt pocket, removed the large thick rubber band with which secured it, and began to remove receipts, notes and other paperwork. I had positioned myself so that I was facing out of the bedroom door and could clearly see the living room entry door. The papers I had in my hand began to accumulate in a small pile on the bedroom rug. Suddenly I could not hear any conversation coming from the living room. At the same instant I felt a cool breeze and noticed there was an outside window left partially open in this room. Strangely, there were no outside sounds coming through the window. The room became enveloped in an eerie silence—a silence so deep, it penetrated the pit of my soul. I peered down the long hallway and noticed the living room door appeared to shimmer with a peculiar dull light. The shimmering seemed like venetian blinds turned vertically (the same effect Glen described when his "visitors" would leave) . I noticed it was not just the door itself that was involved, but also the door casing and the surrounding area extending from the floor to the ceiling. I began to feel nauseated. My mind raced. I thought I was having a stroke and then a strange, almost amusing thought crossed my mind. *If I was indeed having a stroke, what better place to be than sitting in a wheelchair!* At least I would not have far to fall. I reached over with my right hand and began taking my carotid pulse. I counted and looked at the second hand on my watch: "One, two, three, four, etc." My pulse seemed within normal limits, steady and strong. The phenomenon continued and was getting worse. Now the entire hallway was pulsing, like giant blinds opening and closing in unison. The phenomena increased in speed. The silence was deafening. As soon as the pulsing light hit the door frame of the bedroom, I noticed the speed begin to slow. As it slowed, the light became dimmer and soon it was all over—an abrupt STOP, and then a sudden return of all sounds. I was delighted to once again hear the wind, the rustling of leaves, and gentle sounds of birds singing in the outside world. Also I heard the voices of the television crew and could easily make out Glen and Helen's voices.

I slowly proceeded to get out of the wheelchair. I felt a little dizzy with slight vertigo, sat back down, and tried to calm myself. The feeling passed, and I was able to get up once more with no further symptoms. Almost at the exact same moment, the living room door opened and I heard Glen's voice yelling at me to come on in. He told me they had gotten some snacks and invited me to join them. His first remark when I entered the room was that I looked a little pale and perhaps I needed some food.

"Glen, have I got something to tell you," I blurted out, putting my arm around his shoulders. We walked outside in the yard and took a seat on a low planter wall. I told him about my experience in the bedroom and he compared it to what he had seen many times, where his mysterious visitors were seen to operate some kind of apparatus on their tool belts. I knew then what he had seen and now had experienced it first-hand. My scientific mind took hold. "Excuse me," I said and bolted straight up, running to get my abduction detection equipment.

The first finding was an extremely high magnetic field of 8.5 milliGauss in the area of the hallway and some increase at the area of the door leading into the living room to almost 9 milliGauss. I used a radio frequency detector to look for radio frequencies of unknown origin, but none were found. The geiger counter registered only normal background radiation levels. The ultraviolet light showed numerous splatter marks of greenish fluorescence at the door edges and around the door frame. I had checked this area previously and no fluorescence was detected. The hinges holding the door showed magnetic anomalies such as increased field strength and unipolar-like magnetic function.

The only conclusion I could reach was that there had been a non-terrestrial visitation during the filming and the entities involved decided not to show themselves to the film crew or myself. Was this another example of non-terrestrial humor?

In the weeks and months that followed, Glen and his family moved to two other areas of Southern California. He had retired from the Fire Department and had taken another job in the environmental field with a private company. This position only lasted for several months since he had discovered several environment industrial laws had been broken and the company refused to remedy the problem. Glen reported the incident to the State of California, as he was obligated to do, and went into full retirement.

The visitations and abductions did not stop and continue to this day. He has collected more physical evidence such as oily smears found on the bathroom mirror. When analyzed they were shown to contain both inorganic and organic constituents with large amounts of magnesium, copper, aluminum and some rare earth elements.

They did not match any of the 65,000 known substances and compounds in the computer database. In another episode, Glen found magnetic areas on his abdomen and chest. He told me he could stick pieces of ferrous metal onto his skin and they would just "hang there."

His general physician told him there was no known medical reason for this to happen. He was advised to have x-rays taken of these areas, and did so. The x-rays appeared negative for any type of foreign body or diffuse foreign material. I asked Glen to come to my office for an examination and a time was agreed. I asked several members of our Science Advisory Board from A&S Research to be present so they might offer suggestions during the examination. One was a Ph.D. nuclear physicist and the other a specialist in advanced metallurgy.

Glen came at the appointed time and was given a thorough examination. X-rays were taken with and without the segments of metal clinging to his body. It was thought that an x-ray magnetic dispersive pattern might make itself evident. A light-weight paper containing iron filings was placed over the magnetic areas and the dispersal pattern was calculated. After a lengthy examination we came to the following conclusions:

1. Several areas of the anterior thorax and abdomen presented with quarter-sized areas of unknown magnetism with field strength strong enough to attract and support the weight of ferrous metal.

2. Multiple x-rays taken both with and without attached ferrous metal objects demonstrated no positive findings for internal foreign bodies.

3. Multiple x-rays taken with and without attached ferrous objects did not produce findings of magnetic dispersal within the body.

Because of the lack of further findings, this still remains one of the many mysteries involved in the field of alien abduction. I have continued my relationship with Glen and his family and continue to catalogue his physical evidence and experiences.

CHAPTER 4

THE BEGINNING OF THE UNBELIEVABLE

This chapter contains background information and some personal history of abductees who have had implants removed, plus step-by-step details of their surgeries. Not all can be enumerated here, but I will start with Surgery Fourteen.

Each surgical case can be considered special in itself, yet on occasion one or two stand out from the others. This case is one of them. The abductee is male and forty-nine years of age at the time of the surgical procedure. For purposes of anonymity, we'll call him "Jack." By profession, Jack was a military intelligence officer, retired after serving his stint on active duty. Jack was never in actual combat during the six years of his military career and considered himself a mere low-level tech who operated electronic equipment, which allowed him to only carry a "Top Secret" security clearance. He stated that although this was the case, his equipment was installed in places where much higher security clearances were in place. In other words, he was forced into positions where he did not have adequate security clearances to comply with military intelligence law. He also explained he only had limited memories involving UFO's. One of these happened when he was eighteen years old. One day, Jack's father took him and his older brother fishing. This happened to be the same night that his younger brother was born, but his father was not the type of parent who was all that interested in the delivery of babies and, as he directly put it, "with all that blood and gore." Instead, he kept the other children occupied by opting for a fishing expedition. The night was still and the river water they were fishing in flowed smoothly along its banks with barely a ripple. Suddenly their attention was drawn to the sky above and a light traveling from right to left. The object was circular-shaped and very bright as it crossed their field of vision. Jack believed it was very close and estimated it to be only a few hundred yards away, traveling at a low altitude, only a few feet above the treetops. It appeared that smoke or a thick vapor was being emitted from the craft, which gave his father the impression that the craft was having some sort of mechanical difficulty. Jack compared it to a conventional aircraft and noted its aeronautic characteristics were odd as it continued on its flight path. It was then he recalled his father making a joke by telling the two boys, "It sure must be burning cheap fuel." Jack was very curious since he always had an interest in airplanes. He had never seen this type of flying craft before and did not recognize it. He noticed it was very rounded and he could not make out any wing structure. It was definitely not a helicopter. Jack noted his father's lack of surprise, but since his father had worked as a troubleshooter for Sikorsky Aircraft and had served a tour of duty in the U.S. Air Force as a plane mechanic during the Korean War, this led him to believe he had seen this type of craft in the past. Jack recalled his own observations as to its size and shape. He told me it was definitely saucer-shaped with a lighting structure going around the bottom circumference. He had seen the same type of craft in 1999 on a television program called, "Space 1999."

Now grown, initial contact was made with this patient several months prior to the time of the surgery. Jack was compliant and had sent me x-rays revealing a small metallic foreign body on the inside of his right thigh. That area was painful and the military doctors had taken x-rays of the spot in question. Jack denied any history of trauma to the area but told me he instinctively knew that somebody or something had put this in his body. He also explained he had vague memories of being on a non-earthly spaceship and being examined by non-human entities. After receiving the x-rays and having them examined by our radiologists, we determined he had a metallic foreign body on the inner side of his right thigh. We then sent Jack a package containing questions used to determine the physical and psychological aspects of the patient. We also sent a single-page, gradable questionnaire that would determine his probability of actually being an abductee. Once the package was returned to us, the documents were reviewed and cataloged. Jack scored a "42" on the probability test, which is considered in the high probability range.

I have not written much about the health status of abductees who undergo surgeries for the removal of alien implants as they, by and large, all seem to be within the average healthy norm. This case was different, so I will place some emphasis on this fact. Jack's medical history included childhood night terrors, night fears, and sleepwalking. He also indicated he had the ability to call another person into a dream and discuss it with that person after the dream was over. This is very similar to a shared dream. Jack also indicated he had frequent earaches with absence of infectious causes. In addition he had suffered from osteochondritis of the right knee due to Osgood-Schlatter disease, pains in the abdomen and chest of unknown origin, and visual deficiencies which required him to wear glasses when he was in the fifth grade. During his adolescence his earaches subsided, as did the discomfort in his abdomen and chest. He also indicated he suffered only a few instances of the common cold and an occasional flu. At age twelve, Jack underwent surgery for repair of the Osgood-Schlatter disease's damage to his right knee; at age forty-five he underwent a number of dental procedures for removal of amalgam fillings.

Jack's symptoms began with a small painful lump in his right thigh. Shortly thereafter he began to lose weight—seventy pounds over a relatively short period of time in 2003. His concern was expressed to several medical physicians who made a diagnosis of a cyst without taking any x-rays. Jack had doubts as to the credibility of these physicians—and even began to believe these were not real medical doctors at all.

Jack's only history of trauma was his involvement in several auto accidents in his early twenties. He listed his main five fears as:

1. Financial worries.
2. Loss of right brain abilities.
3. Inability to face fears as they arose.
4. Not being able to bungee jump again.
5. Not finding a mate.

He then began to recognize symptoms of fatigue, memory problems, over-ambition, insomnia, and bowel disturbances.

Jack had a rather bland history of drug experimentation when he was young and stated that he did not like the effects of marijuana. He occasionally consumes alcoholic beverages and suffers today from mild diabetes mellitus with diabetic neuropathy.

His interests after retirement have been centered on gardening, researching and reading. His best subjects in school with main interests today are math, science and physics.

Jack's family history was unremarkable with his father living a healthy life and passing at the age of 71. Jack's mother, also 71, was still alive and well at the time of his surgery.

Jack did admit to two periods of missing time, but does not relate either of them to UFO events. He also talked about having scars and marks on his legs that neither he nor his parents could explain. He related dreams of a war going on in the night sky with multiple strange-looking craft fighting each other. He also recounted a number of episodes where he had had OBE's (Out of Body Experiences) without any trouble returning to his body. He had also seen strange balls of light in his bedroom and had episodes of unexplained paralysis which would inhibit his ability to travel outside of his body. At the time of surgery Jack had not undergone any therapy relating to the alien abduction phenomena or any regressive hypnosis to increase his memory retrieval.

Jack was first seen in my office on 12-12-2007. At that time he underwent a complete history and physical examination performed by our general surgeon and myself. The examination revealed a relatively healthy male for his age of forty-nine whose only systemic malady was a mild case of diabetes mellitus. Jack appeared on the lean side and told us of his extreme weight loss and lethargy following his discovery of the object in his right thigh. He related his opinion of how the military had used him in some way as an experiment with the extraterrestrials, insisting that these non-human entities had implanted him with some sort of a device, which caused his weight loss and subsequent health problems. He explained his desire to have the object surgically removed and analyzed.

At this point our testing centered on the object within his right thigh. He had about a dime-sized nodule readily felt below the surface of the skin. It was firm and elevated with more discomfort on lateral pressure than by direct palpation. An ordinary electronic stud-finder was used which pin-pointed a metallic object directly below the area of the lump. A Gauss meter was then used, determining the presence of an electromagnetic field. The Gauss meter registered a field strength of 6 milliGauss. Subsequently, we used a Geiger counter to search for radioactivity, yet none was detected above normal background radiation levels. The next instrument was a radio frequency detector which detected two separate radio frequencies coming from the object within Jack's body: 25.893503 megahertz and 59.125254 kilohertz. In addition, an ultraviolet examination with a three-frequency light revealed no fluorescent marks in the area of the implant.

The following day new X-rays were taken as well as a CAT scan. The radiologist verified the object's size and location. The report isolated a 3mm radio-opaque density in the subcutaneous soft tissues under the skin surface within the subcutaneous fat, representing a calcification or foreign body. No disruption of the skin is noted on the CAT scan. There is minimal amount of stranding of the fat around the radio-opaque density. No abscess or fluid collection is noted. The underlying musculature on the right thigh appears unremarkable. There is noted atherosclerosis vascular calcification.

On the same day the patient underwent pre-operative blood and urine testing, the results evaluated by our surgical team. The findings were consistent for a forty-nine-year-old male with mild diabetes mellitus, and it was determined the surgery could be performed without complications of general health problems.

On the following day, December 15, 2007, Jack underwent surgery for removal of the object suspected of being an alien implant.

THE SURGERY
Our surgery team and television crew was standing by when Jack arrived at the surgical suite. He was introduced to the team and given preliminary instructions as to what was going to take place. After changing into a surgical gown, Jack was escorted to a treatment room, and a blood specimen was taken from his arm. This was then placed into a centrifuge which separated the cellular portion from the liquid portion of the blood. The liquid portion would be used postoperatively to house the surgical specimen.

A diagnostic ultrasound unit was used to visualize the foreign body prior to its removal. The procedure was successful and an image was made of the object within the inner portion of Jack's right thigh.

Jack was escorted into the operating room and placed upon the operating table. He asked what position he would be in and was told he would be on his back with his right knee flexed. Jack told us he wanted to see the entire procedure performed and would prefer to be sitting up so he could get a good view of the surgical site. Both Dr. Matriciano (the general surgeon) and I were amazed at this request, as this was not common. We discussed the matter privately and decided we would grant Jack's wish.

Jack was positioned on the table propped up with several pillows and covered with a draw-sheet, but his right thigh area was visible. A routine surgical prep was then performed on the operative site. The surgical area was subsequently shielded with sterile drapes. A C-Arm X-ray device was used to image the area on two television screens prior to the administration of anesthetic. This was done to insure the operative site was directly above the foreign object. Next, we administered a single injection of local anesthetic to the area. Once done, the general surgeon and I left the room for our surgical scrub.

When we returned to the operating room, the patient was informed the procedure was about to begin. A single incision was made over the suspected implant area and carefully deepened to the level of where we thought the implant should be. The C-Arm was activated so that we could see our instruments in real time within the depths of the tissues and see how close we were to the object. Numerous attempts were made to place a clamp on the object, but each attempt was a failure because, for some unknown reason, the object seemed to move away from the approaching instrument. We did not know whether this was a magnetic effect or something else we did not understand. It is a fact that all surgical instruments are not capable of being magnetized and if this was the case, it added to the mystery of why the object was moving away. It finally took four hands and several instruments to isolate and trap the object so it could be clamped and brought to the surface of the incision. Once that was done, it was severed from its attachments to the surrounding tissues. The object was placed on a sterile surgical sponge and superficially examined by myself, the general surgeon, and one of our scientists from A&S Research. It appeared very similar to other objects we had removed in this category, which were metallic and covered with a biological coating. It was a dark gray in color, measuring about 4 mm in length (with the biological coating) and had the diameter of a thick pencil lead.

The object was then placed in the container of blood serum previously removed from the patient and taken into a small laboratory in the same office. We examined the object superficially under an optical microscope and determined the basic structure was very much the same as some of the previous objects. After this examination, the object was re-introduced into the serum solution and the container sealed and signed by several witnesses and myself. The next time it would be opened would also require several signatures by witnesses, thus insuring the unbroken chain of physical evidence.

The patient had a small bandage applied to the surgical area without discomfort. He was given post-operative instructions and transported back to his hotel. Jack was also seen the same evening at dinner and had no apparent discomfort from the surgery.

Jack was followed postoperatively for several months. Initially he had some bouts of depression which only lasted for a few days. He then rebounded quickly in a phenomenal way. His health returned with vigor and his weight returned to what it was before he entered military service.

Jack was so curious about his abduction relating to the implant that he requested we perform regressive hypnosis before he left California. We contacted Yvonne Smith, noted hypnotherapist who was trained by the late Budd Hopkins, and she agreed to perform the hypnotic procedure. I was present during the session and can attest that the patient was in a deep state of hypnosis. Once the session was over, the patient confided in me his belief that he was "never hypnotized at all." I withheld my observations as I determined it was best for his psyche.

During the session he went into some detail as to being on a craft and in a room where a machine was activated remotely, making a deafening high-pitched whine he described as "ear-splitting," giving him much pain. During this time Jack was seated on a protrusion that jutted out of the wall of the room, on a pedestal, or perhaps on the floor. He could not tell. The machine he described had an arm that came down over his right thigh area and inserted something into his body. *(The complete scientific results of our laboratory testing of these objects can be found elsewhere in the book. Readers with backgrounds in science, metallurgy, or chemistry will find them of particular scientific interest.)* After that, the unbearable sound fully disappeared, his pain dissipated, and he was taken from the room by the typical gray alien beings.

SURGERY NUMBER 15

We will call our fifteenth surgical candidate Steve who, at the time of his surgery, was forty-seven years of age. He was local to the Southern California area and as it turned out, only stone's throw away from an additional office I maintain in Camarillo, California. When Steve made his appointment, he did not indicate to me or my office receptionist that his chief complaint was anything other than a possible foreign body that had gotten into his foot.

I first saw Steve about a week after his original call. He made an interesting impression on me from the beginning—tall, dark-haired, with a pleasant but quiet nature. Steve underwent a thorough lower extremity history and physical examination. He told us he was married with several young children at home and what he described as a happy, well-adjusted home life. His explanation of his toe problem was very precise. He told me that he awoke suddenly one morning with acute pain in the second digit of his left foot. He explained it felt sharp as if something had pierced his skin and he could feel the bulk of the object within the toe. He found a couple of small drops of blood on his bed clothing and on one toe, but he also noticed some small puncture-like marks on the digit itself and what could have been two or three small puncture wounds; however, there were no other signs of injury. I asked him if he saw bugs, spiders, or other insects as I was thinking this might all be the result of a bug bite. Steve denied these possibilities and insisted that he had a metallic foreign object in his toe. In over forty-five years of podiatric practice I had seen and removed hundreds if not thousands of foreign objects from the human foot. From the onset of meeting Steve until the time he left my office, something seemed odd. The case left me with a strange impression. I could not pinpoint the origin of my intuitive thoughts at the time. I mentioned this to my secretary and she agreed that Steve did not react as one of our ordinary foreign-body foot patients. Before he left I gave him a prescription to have x-rays taken of his foot and instructed him to bring the films with him to his next visit. Steve made another appointment for the following week. When I brought his file back to my office and began to fill out his complete history and exam, I was impressed with a simple note that I had made. It was the unusual finding of no noted trauma at a suspected foreign body entrance site. Although I did not know it at the time, I had just taken the first steps into that mysterious realm of "The Twilight Zone."

My second meeting with Steve was the Friday of the following week. His demeanor was more relaxed and he seemed to be more at ease in general. He handed me the x-rays that were taken at a local hospital radiology lab. When he handed me the package a broad smile swept across his face. I carefully pulled them from the standard x-ray envelope and one by one slipped them on to the view box. Even before I had a chance to make any remarks Steve walked over, stood at my side and without hesitation declared, "It's there isn't it?" With that he took one of his long fingers and pointed to the base of the one of the toe bones in the second toe of the left foot. Before answering, I removed the film, carefully turned it at several angles and replaced it back on the view screen. I then told him he was correct about having what appeared to be a metallic foreign body in the second toe of his left foot.

I asked him to have a seat. It took only an instant to notice that he was now 100% more relaxed. I directed my gaze like two sharp arrows towards his eyes and made the following statement: "Okay Steve, now do you want to tell me the real reason you are here seeing me as a medical professional?"

Steve then brought forth his story of how he believed he was an alien abductee and that he had heard me on the radio and seen me on TV many times. He also explained that his employment was not far from my office so he decided he wanted to accomplish two things at the same time. The first one was that he wanted to acquire some first-hand information on my genuine foot specialist procedures and what my approach would be toward a person who would finally confess that he was not an ordinary foot patient but rather an alien abductee with a metallic implant. The second thing, he said, was to discover the truth about the implant.

Steve was so down-to-earth, practical and convinced of his abduction experiences; however, his approach caused suspicion. I wondered how legitimate this implant might be. I could not let myself be led astray by some clever con artist or reporter who might be trying to get me to claim that a common object was an alien implant. I decided to treat Steve as an abduction patient and put him through all the hoops that I had with the other fourteen cases, including a couple of other stress factors that might give him away. I explained to him he would be given a packet of questions to fill out at his leisure and there were no right or wrong answers; he could take as much time as he needed to accomplish this task. I also explained that I would take his x-rays to our own radiologist to get his opinion in reference to the visible foreign object in the second toe of his left foot. I assured Steve that all his information would be held in confidence and only the scientific and related aspects of the case would become public knowledge.

Before Steve departed my office I asked him what his occupation was and learned that he was in the science field and specialized in nano-technology. My next contact with Steve was two weeks later where he made an appointment in my Thousand Oaks office. Since I had the x-rays with me, I asked Steve if he would like to go over them again. He was excited to do this as each time it seemed to re-affirm the reality of his situation. I displayed the films, each next to the other on the view box, and Steve pointed out the area of the foreign body even though I had altered the way the views were displayed on the screen. I would soon take the films to our radiology consultant for his professional opinion.

Steve was then escorted to one of our treatment rooms by my nurse, placed on the examining table, and made ready for a number of tests that only our abductee clients are subjected to. These include an ultraviolet light examination (using varying frequencies) of the involved and surrounding areas. In previous instances we have found fluorescence of varying colors in or around the affected areas. In Steve's case, none were visible. Another test involves a simple carpenter's stud finder which will detect metal when it is under the skin. Steve had one positive area at the top of the second digit of the left foot. A Gauss meter was used to look for magnetic and electromagnetic fields. A tri-meter was also employed and numerous large spikes in excess of five milliGauss were discovered emanating from the area. The next device used during the examination was a radio wave frequency detector. We did not have the use of a radio frequency recorder so we were not able to record the radio waves themselves, but only noted their frequencies in megahertz and gigahertz. These turned out to be the following: 17.68658 gig. and 14.749650 meg.

At that point our patient handed me his completed "abductionaire." I noticed that he had written more material than any of the other abductee surgical candidates. Steve illustrated a complex history of what could be termed a typical alien abduction scenario which started in his childhood and continued through his adult life. He rated them from "mildly upsetting" to "moderately severe."

Steve was born on May 4, 1960, in Evanston, Illinois. His mother had a normal pregnancy, but Steve told us the delivery was difficult. He did not relate the details. During the years of his early childhood Steve suffered night terrors, sleepwalking, and fears of the unknown. He also had a history of the average childhood diseases without complications and was generally in good physical health. He noted one serious accident which occurred when he was twenty-three years of age when he fell out of a boat and injured his neck without permanent pain or obvious deep injury, for which he receives yearly physical examinations from his local physician.

Steve had a negative history of addictive drug use and consumes alcoholic beverages only on social occasions. His answers to most of the abduction questions were very verbose with great attention to detail. He had attended the University of California at Los Angeles and attained his degree in chemistry and material science.

When the question arose asking if other members of his family were also involved with UFOs or actually witnessed events, Steve's answers became voluminous. It would appear that most of his family, including his wife's family, had been consciously involved for many years—and most without intimate communication. Steve marked two pre-printed drawings that were part of the abduction package to indicate where he thought he might have foreign objects or alien devices. One spot that he marked was the second toe of his left foot; the other spots were related to his head. Steve also attached numerous pages of further detailed information pertaining to his encounters—whether they occurred alone or with others.

My next step was to make arrangements for the patient to have a CAT scan of the involved area to help determine the exact size and location of the object and its relative relationship with the underlying bone. I was fortunate the radiologist was available and the facility close to my office. Since the funds for the surgery were being covered by a small television production company, I had to make sure that the crew would be welcomed at the radiological facility. Fortunately, the facility was not busy that day, so everything was a go.

We went over there with the crew—moving all the lights, cameras, and other TV production equipment to a suite in the building next door to my office. After the introductions were made, I asked the radiologist to take a look at the original x-rays. He placed them on the view screen, took about a three-second look and said, "You seem to have a metallic foreign body in the second toe of your left foot?" He asked Steve a few questions and then commented that he must be one of my special cases.

Steve was then escorted into the CAT scan facility and we watched and filmed through the glass enclosure as the scan proceeded. Once the last view had been taken, our surgical candidate was brought back to the viewing room and once again our radiology expert reviewed the CAT scans. He gave us all the details pertaining to the foreign metallic object. It appeared to be close to 6mm in length and was about the diameter of a lead pencil. The foreign object was extremely close to the bone, it had not penetrated the osseous tissue and the radiologist predicted no complications with removing the object surgically.

With this preliminary work being done, we were well on our way to picking a date to do the procedure. One final chore was to send Steve for his routine pre-surgical lab. This would have to be done prior to the surgery in case there were any unexpected complications. I made arrangements for Steve to go directly to a nearby medical lab that could get the results to us in short order. Before Steve left, the television producer wanted to do some extra shots the following morning and requested shooting them in my Thousand Oaks office. Arrangements were made to meet at 9:00 AM.

The following morning the crew arrived a bit early and started setting up the equipment. Soon it was 9:00 AM and our patient was not there. I tried to reassure the crew—and myself—that traffic was bad that time of morning and Steve would surely be there soon. Time began to fly and soon it was 10:00 AM and we had still had not heard from Steve. I tried all his available telephone numbers and was not able to get in touch with him. Our producer was becoming more irritated as the minutes flew by and finally threatened to pack up his crew and equipment. Suddenly my phone rang. It was Steve. What he told us turned our day into an unplanned and unexplained series of events. We listened spellbound as Steve related the story.

Steve had left his house in Fillmore, California in plenty of time to arrive at my office early. He took a route over the very steep Grimes Canyon—complete with hairpin turns and switchbacks. According to Steve, he was often the only car on the road. He drove a fairly new Cadillac which he deemed very dependable, having never had a severe breakdown; however, when he had almost reached the top of the steep canyon grade, the engine just quit. He jumped out of the car, raised the hood, but nothing seemed wrong. There were no apparent loose wires, no broken hoses, and no steam. The battery would easily crank the engine, but the car would not start or run. Since he could get no cell phone reception, he decided to hike back down the canyon to get help. He finally found a tow service that gave him a ride to where he left his car. When the mechanic tried to figure out what the problem was, he came across a small computer module that had been literally fried. The sides were melted, and yet all the surrounding engine and attachment parts seemed untouched. When we heard the news, the TV producer became so excited he told me to tell Steve to stay right where he was and that we would come and pick him up. Steve agreed to the plan and soon we were loaded with all the camera equipment and on our way to meet Steve.

We met Steve at the address he had given us in Fillmore and were able to look at the car and interview the mechanic on camera. Things were apparently as Steve had described on the phone—his computer module had been partially melted without any damage to the car's engine. The mechanic told us he had never seen anything like this before. We left Steve's car at the garage and then made the return trip to Thousand Oaks. On the way back we stopped to film the spot where Steve's car suffered its damage. The intended purpose for the day was to finish shooting scenes we needed for some of the post-production material. Some of this material is called "B-Roll." One of the planned shots was for me to open up Steve's x-ray folder, remove the films, place them on the view screen and have camera operator film them with me talking about the findings.

I kept a pile of x-rays stacked on one area of my desk which represented cases that I was currently working on. I reached into the middle of a pile of about fifteen different x-rays, still in their envelopes, and found Steve's without a problem. I told the director that I had the films in hand and was ready whenever they were. In a few minutes I was told we were ready to shoot the scene. I brought the x-rays into the room, opened the envelope and began putting the films upon the x-ray view screen. What I saw next was something I will never forget as long as I live. It was something without explanation. On the outside of all professional x-ray envelopes is some pre-printed writing indicating the name of the patient, the date, x-ray number, and other information. Also in large letters was a sentence stating these films had already been processed. I can honestly say that in the forty years of my practice I never once paid much attention to what those envelopes had written on the outside. But this writing was suddenly on the *inside*. You can imagine my shock to see the exact duplicate of this printing from the envelope imbedded into Steve's x-ray films. I explained to the crew and producer the unusual nature of what was just witnessed and what a physical impossibility this was. I suggested taking the films and their envelope to the radiological facility next door (where we previously had been for the CAT scan).

The crew grabbed their camera and equipment and we bolted out the door. The radiology lab did us a favor and agreed to look at the films and paper jacket—and also agreed to an interview with the radiologist and techs. As it turned out, we heard the same explanation from them all: *"This was impossible."*

The radiologists and techs explained that once the silver salts on the film had passed through the developing stage, the film was then immersed into a fixative bath. The silver salts left on the film could not be altered. I also explained how I kept these films in the center of a pile of other films and none of them showed the transferred writing. This did nothing but add to a mystery that remains unsolved till this day, along with the strange mechanical damage to Steve's car.

All the preparations for the day had now ended and the surgery was set for the next day at 9:00 AM, which was to take place at the general surgeon's facility. The camera crew was given a call time of 7:00 AM so they would have time get set up prior to the surgery. All volunteers who were invited to participate were also scheduled to be there at 7:00 AM. I knew from past experience that on the surgical day there was never such a thing as too much time—there were always unexpected experiences.

THE DAY OF SURGERY

The day began on two pleasant notes—the first being the weather was absolutely beautiful, and the second was that almost everyone arrived according to schedule. We had invited about fifteen witnesses, including some serious UFO researchers such as Whitley Strieber and his wife Ann. Others were writers, our own A&S Research photo crew, and those who were employed strictly for this event. In order to extract metallic foreign bodies, it is necessary to have a specialized surgical device called an X-ray C-Arm. This is a portable x-ray device that uses low levels of radiation to produce a moving x-ray which appears on one or two television monitors. With this device, the surgeons can see the placement of the surgical instruments inside the open incision which makes it easier to grasp a foreign object and excise it. Steve arrived only slightly late because of heavy traffic, but the crew wasted no time getting down to the business of the day. The first event was a complete medical history and examination by Dr. Matriciano, our general surgeon, which was followed by a blood draw in which we use a centrifuge to spin the whole blood until it is divided into a liquid portion, the serum, and a solid portion which is the cellular content of the blood. The serum is then separated into another container which is used to contain the surgical specimen. This was a special technique developed by A&S Research to protect and preserve the integrity of the suspected implant through the testing period.

We had another piece of equipment made available to us that day which we had not used previously. Basically it was an optical microscope that produced the world's most powerful images through its lenses.

Steve was then escorted to another room where we had a portable recording ultrasonic device similar to what a fetal ultrasound device looks like. The device produced two recorded ultrasonic images of

Steve's second digit containing the metallic foreign body. This was the last test done to prove the object was imbedded into the second toe of the left foot. Our surgical candidate was then escorted into the operating room, positioned on the operating table, and made comfortable. His blood pressure and other vital signs were taken and the surgical area prepped with germicide material. The surgical area was then draped with sterile drapes by our scrub nurse. Dr. Matriciano and myself used the time to scrub for the surgery and change into our scrub suits and gloves. Although this was a foot case, it was decided that our general surgeon would do the major part of the surgery and I would become the assistant. This worked well, as many times I would be called out of the room to help clear up a number of TV technical and/or medical problems. For example, one of the witnesses watching the surgery from the waiting room noticed the audio from the surgery room could not be heard, and I was called out. Fortunately, one of the TV technicians figured out the problem. I returned to the operating room and narrated the proceedings to the viewing audience, making them more familiar with the surgical room and this type of procedure.

Finally, while deeply probing the wound, one of our instruments touched something that sounded metallic. We heard the sound and surveyed the TV screens. After that happened we also witnessed part of the foreign object move out of the way as if to avoid the next touch of our grasping weapon. This happened several times and finally, we felt solid material in the jaws of the instrument and carefully drew it toward the surface of the wound. As we pulled the object into the open air, we were able to see that this was only a small piece of it. It was dark in color, shiny, and smooth. We carefully placed the partial specimen on a sterile drape to be removed later and placed into the blood serum container. Almost the same procedure occurred several times over again—the metallic portion of the specimen was fragmenting as we removed it. Finally, according to our visualization on the C-Arm monitor, we had one remaining metallic portion to remove. We carefully followed the original portal we had surgically made and followed the tip of the instrument until it was pressing against the remaining piece of the foreign object. At that point we had our C-Arm tech push the pedal that would take the photo showing we had almost completed the surgery, verifying we only had one more piece of material to remove.

The pedal was pressed, the pictured formed on the screen almost instantaneously, to all our surprise, the remaining single piece of the object was *not there*. More views were taken, but no further evidence of a foreign body could be found. This was another one of the mysteries that would make this case unique.

The already obtained pieces of specimen were taken into our lab. Each individual piece of metallic and biological material was transferred into the waiting specimen bottle. It was decided this would be the ideal time to look at some of these objects under the high magnification of our new powerful optical microscope. A couple of the small pieces were then mounted onto a specialized optical glass stage and covered with a specially-made glass cover. After a few minutes of mounting the glass slide on the stage and adjusting the variable lighting sources, we began to see an image on the screen. Our curiosity held us paralyzed and frozen in position as were we all looking at things we had seen many times before, but not in such vivid color—with almost 3D imagery and clarity. Suddenly someone shouted out, "What in the hell is that?"

With a pencil, he pointed to an area on the screen. We remained stunned as we watched the object we were viewing physically turn and adjust its position so that we were now staring at another part of the object! After our shock subsided, I called it a day. We thanked everyone for their presence and participation and invited a few to the hypnosis session that was to be performed by Yvonne Smith. Steve was able to stay that night in a nearby hotel so we finalized our plans to meet there for dinner and some conversation.

It was evident at dinner that everyone was more relaxed—as if the pressure had partially lifted. We discussed the day's events and all the mysterious happenings. One of the subjects I have not yet touched upon was Steve's abduction experiences as well as the magnetic and other anomalous findings at his home. Steve's experiences started in early childhood. He believes his parents were also involved, but these early events were never brought out or discussed among the family. His experiences, in general, were typical in character. He was floated out of his room in a beam of light and later returned with little or no memory of what had happened.

When Steve married and began to raise his own family, he became aware that his immediate family was also involved. His wife would not admit to any of her own experiences, including one some years ago when they wee both traveling on a remote highway in the early hours of the morning. A bright, rounded floating object followed them for a mile or so and then closed the distance between them. Nervous, Steve pulled over to the side of the road. The flying craft, which was very bright in color, settled into a hover mode. Paralysis overtook him. He could no longer turn his head or reach for his wife. He now noticed that the open sky was visible through the roof of his car—and his wife was beginning to float out of the car through the open roof. Although he couldn't move his muscles, he remained conscious and remembers being told that his wife would be just fine and would be returned shortly. Though Steve had learned to trust his non-terrestrial captors, their reassurance did not subdue his fears.

Steve had seen a number of different beings through the years, the majority of these being the grays. He described them as not being all from the same culture. The majority of them were short, but there was also a taller species that appeared older and had some indication of sex which became apparent through their tight, form-fitting uniforms. Others of this species were more animalistic with no apparent attire at all.

Steve has also witnessed other non-terrestrial beings, including some that looked exactly like us, yet with more perfection in their features and mannerisms. Steve concluded that all the races of beings with whom he had come in contact with were in some way genetically related. The common method of intercommunication was telepathic; still, he had audibly heard transmissions including strange buzzing and other noises. Steve's relationship with these beings grew through the years: many times he was not taken out of his home at all but had whatever experiment was being done to him carried out right in his bedroom. By experiencing so much contact, his family was not immune. As his children became older, they reported the events of their nighttime experiences to their parents. At this point Steve's wife also began to admit that she was involved.

CHAPTER 5

SURGERY NUMBER 16

Our next case is Surgery 16 and our patient is from Old Hickory, Tennessee, whom we will call "Phil." Although he had seen me on television and heard me on the radio, Phil contacted Budd Hopkins who referred him to me. My first communication with Phil was by telephone at which time he told me that he had a metallic object in his right wrist and that it had probably been there since he was about eleven years old. He told me he had an x-ray that he would send which revealed the metallic foreign body. A few days later, the x-rays arrived. After reviewing the film, I sent him the standard multi-page questionnaire that all of our prospective surgical cases receive. I then made arrangements for one of our radiologists to review the x-ray for me.

Our radiologist verified that there was indeed a small metal object in Phil's right wrist. I notified Phil and asked how he was coming along with the packet of questions, explaining that there was no "right or wrong" answer to any of them. I was pleased to learn that he was almost done and would have it back to me as soon as possible. Several days later the package arrived. I started by grading the Abductee Probability Test: Phil scored a "28" which put him in the low probability range. I read on and was impressed by the large amount of detail—especially a childhood experience that we had not discussed—a camping trip with two friends when he was about eleven or twelve years of age. This was an interesting incident that we'll cover later.

Phil was born in Orlando, Florida, on November 19, 1961. His mother's pregnancy was normal in all aspects. He has two half-brothers and two half-sisters with whom he has positive sibling relationships. He is a high school graduate with hobbies that include an interest in cars, motorcycles, and boats. He is currently busy restoring his collectible Buick. He has never seen a psychologist or a psychiatrist and denies ever having had any psychiatric episodes.

Phil considers himself overly ambitious; he gets headaches that he believes are work or stress-related. He works in the flooring business, which requires the continued use of his hands and wrists. This may have led to the pain he began to experience in his right wrist. Doctors attributed this to carpel tunnel syndrome. Until having an x-ray of the pain-plagued area, he was unaware of the metallic object and no surgery was performed at that time.

I studied Phil's answers. When asked to "List the benefits you hope to derive from this therapy," he replied, "To finally and hopefully understand what this object is in my arm." Phil's answers to questions about UFO involvement have differed from his responses to more specific questions. For example, when his friend "Don" recalled seeing a craft the size of a football field, Phil denied seeing it. However, when asked, "Have you ever seen a UFO?" he replied, "When I was twelve or thirteen years old I saw three bright blue lights in the sky that formed a triangle." And that brings us back to the camping incident.

When about eleven years old, Phil and his buddies decided to camp overnight in a large rural piece of property not far from Phil's house. Excitedly, they gathered blankets (for tent-making of their own design), sleeping bags, drinking water, and food to cook over a campfire.

Upon arriving on the property, the three kids scouted out the best site for the tent and the perfect spot to see stars—a necessary element for the ambience of telling scary stories. Neither Phil nor his friend Don could recall exactly what time they decided to turn in that night. After retiring, sometime in the wee hours of the morning, Phil became restless and fidgety. He remembers opening his eyes to find the brightest star he'd ever seen right over the campsite—it was so bright that he thought it worthy of waking his two companions. They fought his efforts to wake them, but once roused, the three huddled together, watching in awe as the star became even brighter and started to descend right over them, becoming bigger and bigger as it drew closer and closer. Phil and his friends were not afraid at the time, but rather filled with curiosity as to what the bright object might be. The next thing the boys knew, they were floating in the air inside a large beam of light, heading towards the bottom of the object. They weren't afraid, but excited about flying through the air. They were having the time of their lives—this was FUN!

I have contemplated how I would feel if I found myself floating through the air in a beam of light—I'd be frightened to death knowing that at any time I could fall to the ground! I've come to the realization that children have totally different concepts of what they perceive as fun.

The three were approaching the bottom of the craft where they could see an opening. They floated through that aperture into a round room. It was then that the boys' feelings changed from that of fun and excitement to mild fear mixed with curiosity.

In Phil's questionnaire he stated that during early childhood he suffered from night terrors and nail-biting, but otherwise considered his childhood a happy one. His state of physical health was normal during both childhood and adolescence, with no major physical illnesses. However, when Phil was around twenty-five he was involved in an auto accident, but denied injuring his right wrist, hand or arm at this time. He also reported that when he was twenty-five or twenty-six, while staying over at a friend's house he had gone to bed for the night when he felt a presence in the room—and found himself paralyzed. He stated it felt like someone or something was holding him down and he didn't even have the ability to scream. It is not surprising that Phil lists his most major fear as "being alone."

After reviewing the x-rays and examining his Alien Abduction Questionnaire, my team and I determined that Phil was a candidate for removal of the object from his right wrist. The problem at that time was that there were no funds to complete the project. Several organizations and individuals offered the necessary funds, but none of the offers were enough for the immediate time frame.

As I continued conversing with Phil and his wife by telephone, we developed a rapport. He became more than just another "abductee surgical candidate" who was looking for answers. So, while we were waiting for funding, I thought that we could put our extra time to good use by attempting to gather further data. I agreed to send Phil a couple of the instruments we use during our preoperative examinations. One was a Gauss meter, and the other a radio frequency detector. After Phil received the instruments, he called and told me that he found a large magnetic field surrounding the object and also radio frequencies in both the megahertz and gigahertz bands: 137.72926 MHz and 516.812 GHz. I was excited to hear this news and told Phil I would be back in touch with him as soon as we acquired financing. Several more weeks went by and another individual with a sizable organization expressed interest in financing the procedure and asked for an estimate of the costs. I supplied him with the details and he got back to me within a few days, agreeing to finance the surgical procedure and a couple of lab tests. He told me to proceed with the arrangements and we agreed on a date.

I contacted Phil and told him what was going on, and he agreed to clear his schedule for that time. With that, I made arrangements with our general surgeon and the surgical facility. I also called the supplier of the C-Arm (an x-ray image intensifier), and cleared the date with him. Everything was in place to perform the procedure—thus, I was in shock when my guaranteed financial backer reneged, saying that his advisory board had decided the expenditure was too large and they would have to pass on the deal. One of his counter-offers was that I would fly to Arizona and do the surgery free at a facility of his choice and he would pay for the airfare of the patient and myself. It was all I could do to control my next few remarks! My short and well-meant answer was, *"No, thank you. Have a nice day!"*

I then set about notifying all parties concerned. Phil, too, was disappointed. Since there was suddenly no foreseeable date for the surgery, I asked Phil to ship back my equipment. He agreed and added that he had used the radio frequency detector several times throughout the past weeks and could no longer detect any emissions coming from the implanted device. This only added to my discouragement. All efforts in getting the surgery done were speedily slipping away. The cancellation destroyed a major undertaking as all involved (except the surgeon and C-Arm supplier), were volunteers who planned their time well in advance.

Within the next few weeks I found myself in the throes of despair over this case and then, suddenly, a miracle happened! I received a phone call from my good friend in Mexico, Jaime Maussan. He had called to chat and catch up on what was new in the world of "implant removal." I explained what had recently occurred and his reaction boosted my mood by 1000 per cent! He asked if I had someone who was going to film the procedure for professional use. I told him that we only use video footage for scientific archival purposes. With that, he offered to pay for all the expenses—including laboratory tests. I was elated! We bounced around some possible dates, and I agreed to contact all parties involved and get right back with him.

Working feverishly, I was able to set everything up again in two days' time and Jaime, true to his word, sent the check post-haste! Not only did he pay for the surgical procedures, but also transportation costs for Phil, his wife Mary, and his childhood friend Don, who had been on that camping trip and who had some conscious memories of seeing the craft.

So, Phil, Mary, and his friend Don arrived on April 22, 2010. With cameras rolling, we had our first face-to-face meeting at my office. Also present was our chief scientist from A&S Research, Steve Colbern. The patient was made comfortable in an examining room at my office, and I started by taking a detailed history of Phil's background and experiences.

Once that was finished, we used our instrumentation to detect the metallic implant. Afterward, with the help of the radio frequency detector, and to our surprise, we discovered the implant was now emitting radio frequencies in both the gigahertz and megahertz ranges. I researched my previous data and discovered the frequencies that Phil had given to me via telephone were exactly the same as the ones we had just detected. This led us to the conclusion that these implant devices probably have a switching mechanism that can turn themselves on or off. This could be something internal in the device or perhaps a signal from a source far away.

Next, we used the multi-frequency ultraviolet light and were amazed to discover a chevron-shaped symbol on the outside of Phil's right arm. It was approximately two and one-half inches across and was not at all visible in ordinary lighting. This came as quite a surprise to Phil and the others present. Further examinations confirmed the metallic object was in the right wrist, and a CAT scan was ordered for the following day. Jaime had his crew break down the camera equipment and lights, and we all adjourned to the hotel where Phil and his group were staying. Since Jaime was staying there, too, we all agreed to meet for dinner and light conversation.

At 10:00 AM the following morning we all convened at my office. Jaime, the camera crew, Phil, his wife, Don and I then walked to the building next door, which housed the x-ray facility. The CAT scan was performed and the films were read by a radiologist who provided the exact measurements of the object and its relationship to the surrounding anatomy. This information is vital to the successful surgical removal. Jaime and the TV crew used the remainder of that day for interviews with me, the patient and his wife, as well as his friend. I also reviewed the plan for the next day, April 24th. We were to meet at the general surgeon's office at 9:00 AM, which was only about fifteen minutes from my office. Phil, Mary, and Don were to be transported from the hotel to the surgery site by Jaime and his crew.

I was overcome with a multitude of emotions. I was delighted that after all this time our surgical candidate was actually going to have the procedure performed. I was also excited about our unexpected guest who was with Phil at the campsite and underwent the experience with him. Jaime's presence and cooperation was an added blessing.

My mind was racing with the implications that a successful surgery might reveal new data we could disseminate to the world—data that would shock the scientific minds into finally funding projects related to the UFO subject. This subject is affecting all of mankind and will continue to do so in the future.

I left the group and headed home for a good night's sleep in preparation for the eventful day that lay ahead.

APRIL 24TH 2012

Finally the day had arrived. After months of waiting and aggravation, the person with a possible non-terrestrial implant would have his wish fulfilled and have it removed. I was happy to see that upon my arrival, Jaime, his crew, and the patient with his wife and friend were all waiting with their vehicles in the parking lot. They were already in the process of unloading the camera equipment needed to film the operation. I then saw our general surgeon Dr. John Matriciano coming across the overhead walkway that connects the hospital proper to the office building, which contained his own private surgical suite. I instructed everyone to take the elevator and meet me at the back door to the surgical office.

Soon the area became a beehive of activity. Also starting to arrive were some of our invited guests such as Whitley Strieber and his wife Anne, and my secretary and office manager with her boxes of needed materials and paperwork. All witnesses were required to sign a document that they were present for the surgery. This was in keeping with the previous protocols we had set up for the other fifteen surgeries. Cameras were set up in the operating room and a monitor was connected in the waiting area so those who could not come into the surgical area could still see every detail of the procedure.

The C-Arm X-ray device arrived and I told the personnel who brought it where to set it up. They brought the equipment into the surgical suite and began to set it up and calibrate the machine. This particular model had two television screens and that would make it a lot easier for us to visualize the object.

At this point Phil was taken into an adjoining examining room and Dr. Matriciano proceeded with a general health history and examination, including a review of my own records. Dr. Matriciano had also requested an ultrasonic examining machine with hopes of being able to visualize the object via ultrasound. I wholeheartedly concurred, as there is no such thing as "too much information."

We brought the patient into another adjoining examining room where the ultrasound unit was housed. The nurse arrived and retrieved a container of ultrasound gel which would act as a conducting medium between the ultrasonic head and the skin. Soon we were underway, studying the monitor, trying to locate the object. Suddenly, we shouted simultaneously, "There it is! I think I see it!"

Indeed, we were right over the object and it could easily be seen on the ultrasound monitor. "Let's get a couple of photos of that," Dr. Matriciano stated excitedly, and with the push of a foot pedal, a picture began to generate from the printer connected to the machine. We reviewed the ultra-sonograms and compared them to the x-rays and CAT scans. Everything fell into place. Each part of the puzzle fit and we were ready to commence with the surgery.

Phil was escorted into the operating room and positioned on the operating table. I left the area to change into my surgical scrubs and see if there were any last- minute items that required my services. Once satisfied that everything was going according to plan, I changed into my greens including surgical cap and a mask, which I left dangling loosely about my neck. I returned to the operating room. The patient was on the table and the surgical area prepped and draped with sterile drapes. Dr. Matriciano was busy opening sterile instrument packages, gauze sponges, and other items we would need. The nurse assisted me in donning the sterile surgical gown, and then applied my sterile surgical gloves.

Dr. Matriciano grasped the syringe containing the local anesthesia and advised the patient he would probably feel a slight stinging sensation, then he punctured the skin at the surgical site. The patient winced and I asked him if he was doing okay. He indicated it only hurt for a second and he was feeling no discomfort.

Next, Dr. Matriciano used a surgical blade to make a small incision directly over the area he had marked. With the tip of the blade, he pressed into the opening, and a small clamp-like instrument was inserted to widen and deepen the area to what we assumed was the depth of the implant device.

Once the probing instrument was inserted into the wound, the C-Arm X-ray device was activated and we could see the metallic object, as well as the probe, on the two television screens. The technician who was operating the device was instructed to capture a still photo at the moment it became visible. Now we had the first visual recording of the implant with the surgical instrument. Further probing of the wound did not give us the visualization needed for the extraction. Another problem was whenever we approached the metallic object with the surgical instrument, it would magically move out of the way. This made it very difficult to grasp the object in the jaws of the clamp. We had seen this happen many times before, so it came as no surprise. Finally, after inserting a second clamp into the wound, we were able to manipulate the foreign object into a position where it could be grasped, and we twisted it towards the surface. Dr. Matriciano calmly stated in a louder voice, "I think we can see it now!" I agreed with a hearty, "Yes, there it is. Now all we have to do is free it from the surrounding tissue."

Soon the object was free and clear from its human container and placed on a four-by-four inch sterile gauze sponge. Cameras began to flash. I held up the gauze square containing the object. Soon a myriad of still photos were taken as well as numerous angles captured by the video cameras. While Dr. Matriciano was sewing the wound shut with small nylon thread, I left the operating area with our newly-delivered implant device still in the gauze sponge. We then viewed the object under a microscope. Next, we removed the outer coating of soft tissue and placed the implant into a container of the patient's own blood serum.

Jaime then began postoperative interviews with the patient, his wife, his friend Don, some of our other guests who were witnesses to the procedure, and me. With the major event over, we ordered in food, ate and talked about the surgery. Jaime was quite excited and had plans to make this event into an entire television program for viewing in the United States on the Telemundo Channel, as well as in other Spanish-speaking areas of the world.

With the surgical procedure now done, we reconvened back at the hotel. Phil was elated, feeling as if he hadn't undergone any surgical trauma whatsoever. Still wanting to push forward, he suggested that he have his first hypnosis session following dinner.

For many years I did not include hypnotic regressions in my writings, but after sixteen cases, I've found that useful data is added that would otherwise be lost. Since the complexity of this case is like no other we have previously encountered, almost every detail becomes important. What makes this case unique is that there were three separate individuals involved—all of whom had bits and pieces of memory. We not only wanted Phil to undergo hypnotic regression, but also wanted to extend the opportunity to his friend Don, as well.

For quite some time A&S Research has worked with noted, qualified hypnotherapists in the field of Ufology such as Budd Hopkins, David Jacobs, Yvonne Smith, and John Carpenter. In this case our choice was Yvonne Smith, who became familiar with the case while the patient was being worked up.

Even though Phil wanted to move forward, Yvonne was not terribly enthusiastic about the idea and wanted to wait until the next day. However, she agreed to Phil's wishes and a few of us were invited to attend. We did not want Phil's camping buddy Don there because

Yvonne had scheduled him for a regression of his own the following day. Yvonne did not want to influence what he might have to say should he hear Phil.

We adjourned to Yvonne's room, which she had set up for the hypnosis session. Jaime was also there with his crew and set up the equipment to record both picture and sound. This was more of a challenge than I had considered—the room's lighting had to be dim, yet light enough to be able to film the client. In addition, the ambient sound in the room had to be very low because Phil would be talking in a subdued voice. Still, there was always the possibility that he might have a period of excitement where a large volume of sound might be produced. These were all technical aspects not in my field of expertise.

Once the preliminaries were done, Phil was asked to make himself comfortable on the bed. Yvonne started by explaining to him that he would be in no danger at any time and made sure that he would be able to find his psychological "sweet and safe zone" which he could enter any time he felt threatened or uncomfortable. I have witnessed numerous hypnotic sessions and have always been impressed by the professional manner in which Yvonne handles her clients.

After the preliminary hypnotic induction phase was complete and Yvonne was satisfied with the level of the trance state, she began taking Phil back to the day of the camping trip. Phil appeared to be deep in the hypnotic state, but could only remember a portion of what had happened that night. He remembered being with his friends and seeing the star descend and the light growing brighter and brighter. After that point he did not recall anything of the event until all three boys were completely awake. It seemed no matter what trick Yvonne used hypnotically, she could not get through this wall of amnesia. Just one detail he did remember was that after the incident all three boys had a sudden craving for raw vegetables.

Yvonne then began to bring the client out of his relaxed hypnotic state, giving him a post-hypnotic suggestion that he would be comfortable, relaxed, and get a good night's sleep. She explained to us that in many cases the amnesia state induced by the "visitors" is so strong that it could take several sessions before she could get past it.

Another session with Phil was scheduled for the following day, as well as the session with Don. Perhaps Don could fill in some of the blank spots that were not obtainable from Phil.

We thanked everyone for their hard work and made arrangements to meet in the morning for breakfast. I have always found it strange that when one is occupied, time seems to fly by so quickly—and so it was with this!

After breakfast we were again in Yvonne's room with cameras set to capture the next series of events. Phil was first and was soon in a deep hypnotic trance. Yvonne began asking him details of his adventure aboard the ship. It was like someone had removed a great cloud from his memory and the information began to pour forth. I will only cover a small amount of material as the case is still under research. Yvonne will also be coming out with some of this information in her next book, *Coronado Intrusion: The President, The Secret Service and Alien Abduction.*

Once the amnesia had been penetrated, Phil remembered that he and his two friends were in the presence of stereotypical grays in a large, round, hangar-like room that had other forms of flying craft inside. He voiced concern to his captors—not for himself, but for his friends. The gray beings reassured him telepathically that all the boys would be fine and they would be returned to their own environment soon. Two of the other non-human intelligences took each of the other boys and started to walk away from Phil. He became panicky and once again needed reassurance that his friends would be okay. With that, he was escorted out of the hangar-like facility and down a curved, dimly-lit passageway from which he could see a multitude of rooms. Yvonne asked him what he was able to see in these chambers and he told her that they all looked like small hospital rooms. He was then escorted into one of these cubicles and instructed to climb onto a table that appeared to jut out of the floor on a single pedestal. Phil stated that the table was extremely comfortable and form-fitting, like it was made for him. He noted that the air he was breathing while on the craft was humid and stuffy and although he was a bit apprehensive, he still felt internally that he was part of a wonderful adventure and hoped in a strange way that it would not stop.

Two of the gray beings approached him. One got very close to his face with one huge black eye almost pressed against his skin. He immediately began to feel relaxed and noted that the other gray was using several instruments to examine him. They looked in his nose, mouth, ears, rectum and penis. His eyes were examined by some form of a visual matrix in which he could see a myriad of lines, rectangles and other geometric figures. A device appeared, coming from overhead, and was placed next to his right wrist. He felt no pain, and although he consciously wanted to move his arm and wrist, he found it stuck to the table as if it were glued there. He was then told to get off the table and rejoin his friends. He felt both happy and sad that the experience was over, but looked forward to discussing it with his friends. Phil was then escorted by a single gray entity back to the original room where he had found himself upon entering the craft. The boys were excited to see each other, but strangely, no one talked about the experience.

Soon they found themselves floating once again in the beam of light towards the ground and campsite. Once on the ground, they looked up and watched in amazement as the bright light above them shot up and soon looked like just another star in the heavens. Phil once again described a lusting hunger for fresh vegetables. Young Don told them of a nearby farmer's garden where they could find fresh growing vegetables. They followed him at a run until they reached the garden. They tore the vegetables from the ground and from their stalks, feasting upon them as if they were the greatest culinary delight they had ever experienced. Yvonne questioned Phil, asking if they had ever done such a thing before and Phil denied it wholeheartedly. He added that raw vegetables were not his favorite food when he was that age. *(It should also be noted that this incident is unique in the abduction scenario.)* With that, Yvonne ended the session and allowed Phil to recuperate from his alternate state of consciousness.

Don, our second hypnotic candidate, was made ready for his session and Yvonne used the same induction procedure as she had with Phil. Don appeared more apprehensive and seemed to have a more difficult time reaching a relaxed state. Yvonne used a number of accepted hypnotic induction techniques, but Don continued to resist. Almost an hour had gone by before Yvonne was satisfied that her subject had reached a light trance state. Following this, she spent at least another twenty minutes trying to increase the depth of the hypnotic state. She did manage to regress Don back to the time of the camping event involving the three boys. Don was able to see the bright descending star quite clearly as it grew in size and brightness, all the time descending over their campsite, lower and lower, brighter and brighter. She asked him about details of the undersurface of the craft. He described it as a shiny metal, but could not make out any seams or peripheral structures—no landing apparatus, panels, screws or rivets. He described it by telling us that the metal looked like it was "grown" or "molded."

Yvonne tried numerous times to coax out more detail of the craft and what happened to Don, but soon realized that her subject also had an amnesia block which would be impossible to break through in one session. So she brought Don back to a fully conscious state. We all took turns questioning him, and I personally thought that although there was no breakthrough, his memory was sharper and he could eventually explain more than what he had originally remembered.

We finally adjourned for the evening and made plans to go to Tennessee and try the hypnotic session with Don again, as well as contact the other boy, now an adult. It would be interesting to find out if he, too, had an implant and perhaps have him undergo a hypnotic session if he wished to do so. At the time of this writing, the trip to Tennessee has not yet been made; however, we still look forward to going there in the future.

CHAPTER 6

THE LATEST FROM VARGINHA, BRAZIL

My travels around the world have been extensive. I have gone from Canada to the tip of South America; from the United States to the Middle East. I have talked about UFO's with individuals from all over the planet. When the geography of the United States was involved, questions always arose as to the exact location of events. This always requires in-depth explanation so people can understand where you are talking about. Even the mention of famous places such as the Grand Canyon, Niagara Falls, Yosemite Park, the Mississippi River or the Washington Monument sometimes require further geographical explanations. However, when speaking of Roswell, New Mexico, no such explanation was needed.

Most everyone accepted the fact that this was one place in the world where a non-terrestrial vehicle or vehicles had crashed in 1947. One has to realize that this was a long time ago and the event does not appear in virtually any history book about the United States. So why did this event make such an impression on a worldwide basis? Was it the denials of the military, the print media, and a government campaign to suppress information that actually backfired? Despite these efforts, Roswell was destined for unwanted fame, local ridicule, and an organized cover-up. In my opinion, the truth will never be revealed by the U.S. government. If such disclosure is made to the world-at-large, it will come through a government agency outside the United States or from the huge non-military industrial complex.

Have there ever been any similar events that would give credibility toward the events at Roswell? The answer is yes. Let's look at a more recent but very similar event that occurred in a small community in northern Brazil in January of 1996. The city is Varginha, which still has approximately 130,000 inhabitants. Its industry is mainly agriculture; its people are simple and hard-working. This humble community rocketed to fame in 1996 when the world news reported the crash of a non-terrestrial vehicle within the geographical boundaries of the city.

This event became known worldwide as the "Brazilian Roswell." Some international television programs revealed details of the investigation and the role of the Brazilian military. In addition, a few in-depth books were written on the subject, including one I had written myself entitled *UFO Crash in Brazil*, also in French as *Des Extraterrestres Capturés à Varginha au Brésil.*

This January, 1996 incident not only involved the crash of a non-terrestrial vehicle, but also included accounts of several extraterrestrial beings that either survived the crash or were able to escape prior to the vehicle's impact in a farmer's field. There was no lack of witnesses to the events that followed, including the aforementioned Brazilian military which included the Brazilian Army, the Fire Department, and the Military Police—all separate divisions within their armed forces, and all of them responded separately to the incident.

Two well-respected civilians investigated this case: Ubirajara Rodriguez, a prominent defense attorney for large private and government businesses, and Vitorio Pacaccini, a reporter and writer with the Centro Brasileiro de Pesquisa de Discos Voadores (CBPDV), who happened to be a consultant for *UFO Magazine* in Brazil (in many countries outside of the U.S., the UFO subject is followed more openly and seriously by the media, government agencies and the public). In many ways, Vitorio played a larger investigative role since he was first on the scene as events started to unfold; Rodriguez was not in town at the time of the actual crash.

The events took place as follows. Three young girls who were employed as maids were on their way home from work when they came across a kneeling, seemingly intelligent creature that they described as being brown in color with an oily skin, no hair on its body or head, and having large red eyes. They also observed three prominent protuberances on the top of its head that ran from the front to the back of the skull. Feeling as though they had "just seen the Devil," they ran home as fast as their legs would carry them. Once they arrived, they told their mother of the frightening experience. She did not have a car to drive to the spot, but they finally succeeded in getting a neighbor to take them back to the scene. By that time, the creature was gone.

It didn't take long before this small farming community was inundated, not only with the story of the crash itself, but also of other activities that occurred in the days following. One such incident involved the capture of an injured E.T. by two young, healthy military policemen. One was a fellow by the name of Marco Eli Chereze, twenty-five years old, and his partner who were traveling in a small military vehicle when they saw a strange-looking being attempting to cross a street—a street not far from the site where the three girls had witnessed a creature. They came to an immediate halt. Marco cautiously exited his vehicle and carefully approached the being that appeared to have an injured leg. He put his arm around the creature and guided it toward their waiting vehicle. With only two seats, Marco had no alternative but to place the being on his lap, though neither he nor his fellow policeman were wearing protective garb. They transported the creature to a small medical outpost. No one there could help their "passenger." Next, they took the being to a local hospital where it was accepted for treatment. At the time, a large contingent of military personnel was already on the premises because any accident or medical emergency at the military base was treated at this same hospital.

Upon examination, it was determined that the creature had sustained a fracture to its leg and would require surgery. I came to Brazil shortly thereafter to do a thorough investigation of the events of this case, and did an in-depth interview with the orthopedic surgeon who performed the surgery. I have covered this in some detail in my book. The important point to remember here is that the doctor was freely forthcoming with the details, which involved specific medical knowledge.

Approximately three weeks following his exposure to the non-human creature, Marco Eli Chereze, the young and healthy military policeman who aided the injured E.T., succumbed to an infection that was similar to an Ebola virus. He was buried in secret. His widow, whom I interviewed, was not provided with a death certificate or autopsy report. She also reported receiving no financial assistance from the Brazilian military or government to help her and her young children survive after the death of her husband.

Almost one year later I was at last permitted to review the hospital records and autopsy report on Marco's death. It was apparent to me that Mr. Cherize had contracted some foreign organism from the E.T. and his body's immune system could not ward off its devastating fatal effect.

Another matter of significance in this case is the United States' involvement with the Brazilian government. The Brazilian Space Agency was notified by N.O.R.A.D. (*North American Aerospace Defense Command*) hours in advance that a craft was coming down from space. They gave the agency all the data as to where the craft was headed and the most-likely impact site. The Brazilian Army was well-prepared for the event. Also, at the culmination of the incident, many witnesses watched as the crash material was loaded onto a U.S. Air Force cargo plane in San Paulo. Although the destination was unknown, the assumption can be made that it was headed for a military installation in the United States.

After my initial investigation of the Varginha case, I had the opportunity to visit the area two more times, the last one being in 2007. This time I traveled by car, accompanied by my dear friend Paul, also from the U.S., who had a deep interest in the case, as well as A.J. Gevaerd, noted and respected Brazilian UFO researcher who is also the owner and editor of *UFO Magazine* in Brazil—the oldest existing magazine on this subject in the world, with nearly thirty years of respected publication.

The trip by automobile gave me a completely different perspective on the geographic location of Varginha, which is located in the state of Minas Gerais. I was able to see how the terrain changed from lush vegetation to sparse barren land. Some of the vegetation, when it existed, included sugar plantations. I also learned that this state supplied most of the precious stones found in Brazil. The land was covered with many mines of different types and was as desolate and colorless as if one were looking at a moonscape. Along the way we stopped for food at many small, interesting roadside stands. These establishments sold a form of liquor made from sugar cane—and I can assure you that it's one of the most potent alcoholic beverages in existence!

Perhaps it was my keen interest in Brazil—or just the interesting conversation that made the time fly by so quickly. Soon we found ourselves at the outskirts of Varginha with a multitude of cattle ranches, dairies, and other farming ventures. Since A.J. knew the area, he guided us to a hotel where we checked in. The place was not fancy, but served our needs and was close to the downtown area of the city. After settling in we took advantage of the hotel restaurant with a quick lunch; we then grabbed our camera gear and headed downtown.

Paul had never been to Brazil before so we wanted to show him some of the local businesses. We found a central place to park and headed by foot to the shopping areas. There were several visual surprises since I had been there last. One of them was a huge metallic saucer-like structure that had been erected centrally in the downtown area. I asked about the purpose of such a structure. A.J. said that this building was the beginning of a UFO museum that would be used to present the facts of the 1996 crash at Varginha to the public.

One of my other surprises pertained to some trinkets I had planned to bring home with me. The last time I was in Varginha mostly any curio or gift shop had carved or cast statuary of the alien beings that had landed there in 1996. Now there were none. After asking several shopkeepers, we were finally steered to one small store that carried a few of the alien statues. Since the price was still reasonable, I purchased them all.

Our next stop was at the Brazilian fire station that handled the capture of some of the creatures in 1996. A.J. was hesitant to stop and go in because he had heard that the fire station personnel were warned not to talk about the case anymore—especially with visitors to the community. We found a parking spot directly across the street, rolled our windows down, and began filming and narrating. With that, one of the firemen came out onto the driveway and motioned for us to come over.

A.J. was extremely surprised—suspicious that we could be in trouble. I insisted we follow his instructions, so I opened the door and, with my companions, crossed the street. Since A.J. was the only one who spoke Portuguese, he was the first to greet the fireman, who appeared friendly and invited us into the station. He then introduced us to his colleagues and, with pride, showed us the fire truck that was used during the capture of the alien creatures. It was in pristine condition and appeared just as I had seen it previously. They also explained that they were not the same crew that was there in 1996. Those firemen had all been transferred to other fire stations throughout Brazil within the six months following the 1996 incident; however, some of the "new" firemen were friends with the original crew and still kept in touch with them. That gave us hope for future investigations. After a photo session, we thanked them for their hospitality and headed back to our vehicle.

Our next step was to try to contact the orthopedist who operated on the alien being. At about 3:00 P.M. we called his office. A.J. told his secretary who we were, that we were in the area, and that we would like to pay a visit to the doctor. She placed us on hold and, after about five minutes, returned to let us know that although the doctor was quite busy, he had agreed to see us. Fifteen minutes later we arrived at his office and went inside. Upon entering, it was obvious that he was very busy by the extraordinary number of patients who were waiting. After telling his staff who we were, we took a standing position and waited for three vacant seats. Once seated, our attention was drawn mainly to the clock. The minutes slowly turned into hours. We waited until we were the only ones in the waiting area. This was in clear contrast to our last visit when he rushed out from his private office to greet us and treated us like royalty.

At last the doctor appeared, wearing a typical long white physician's coat. He shook hands with us and asked us to follow him into his private office. (This was the same room we had occupied when I did my original interview with Ubirajara Rodriguez, one of the investigators, and his son.) I noted a distinct difference in the doctor's demeanor—he would not look at us directly. Even though he avoided eye contact, he was pleasant and asked us to have a seat. He took a seat behind a desk, picked up a pen and slowly began to fiddle with it. The more nervous and upset he became, the more he twirled the pen. As we got further into the conversation, he frantically fidgeted with the pen, anxiously moving it back and forth in both hands—periodically throwing it upon his desk.

Since my command of the Portuguese language is miniscule, and we were not discussing in medical terminology, most of the English translation came from A.J. The first utterance from the doctor's mouth was the most shocking. I felt like I was going to fall over in my chair! His initial statements had to do directly with his surgery on the extraterrestrial being. Gazing down at his desk, the doctor did an about-face on his previous testimony, telling us now that everything he had said in his initial interview never happened and the entire story was based on rumors. Again, feverishly fidgeting with his pen, his face wet with perspiration, he continued in this same vein and demeanor for quite some time, pausing periodically so that A.J. could translate for us. Here, in this 2007 discourse, he flatly denied the existence of any non-terrestrial entity ever being in that hospital. When we asked him about other facts of the case, such as the presence of the Brazilian army in large numbers, he explained that away, saying there were always base accidents that were brought to that facility for treatment or surgery. His next move also came as a shock and surprise. Suddenly, with tears welling in the corners of his eyes, he turned directly to me, at last meeting my eyes for a poignant moment. Without changing his cloudy gaze, he addressed his next statement directly at A.J. "I know the doctor from the United States would like to know more about the blood, bone, and other medical and scientific aspects of the non-human anatomy, but unfortunately I can't offer any further information. As I indicated to you, it was all based on rumors and a collection of stories." With that, he gently placed his pen on his desk and rose from his chair. Reaching across his desk, he shook my hand with great emotion. He then turned to A.J. and Paul, telling them in Portuguese that he was sorry if we thought our time wasted with his visit. We left his office in silence, and once we arrived back at our car, we sat there for a good solid minute, staring at each other in disbelief. In trying to decide our next step regarding the doctor, we came to the mutual conclusion that this man must have been intimidated by someone or something that terrified him. We proposed the idea of calling him again to invite him to dinner. It was I who suggested that we have no further contact with him because it was possible he may have been in fear for his life or the lives of his family. I firmly stated, "Let's leave this poor man alone and get the hell out of Varginha!" We all agreed and left, heading back to San Paulo.

So, what really happened in the small agricultural town of Varginha, Brazil? Certainly not much has appeared in the United States news media, print or electronic. What about the internet and the rest of the world? Did we hear any news coming from Japan, Europe, Australia, the U.K., Ireland, or Australia? How about in France with those that have followed the case from its 1996 inception? What I am about to reveal is going to take you beyond your spectrum of belief.

A seemingly unrelated event inserted itself into my investigation of the UFO crash in Varginha. Through A.J. Geveard of *UFO Magazine* in Brazil, several of the most respected Brazilian researchers were presented with an invitation to meet face-to-face with military officers of the Brazilian Air Force. This meeting generated some publicity, but was lost in the media frenzy of other countries that were releasing their own information around the same time.

Astounded, I learned that the Brazilian Air Force was involved in "investigations" in the State of Minas Gerais during the same period that the Varginha incident happened. This was surprising because all military activity in the area of Varginha was handled by the Brazilian military which, as previously stated, included the Brazilian Army, the Fire Department, and the Military Police—all separate divisions and did not include the Air Force. The documents released, however, showed drawings by military pilots, radar operators, and ground personnel whose job it was to provide surveillance of the Varginha area for weeks at a time.

Several months ago, A.J. Gevaerd was invited to a second meeting of the Brazilian Air Force at which over 2000 documents were released which demonstrated amazing details collected by these flying military professionals. To this day we do not know the extent of what the air force discovered; however, in a recent communication with A.J., I was told that more documents will be released in the future—with the added suggestion of a joint effort of civilian researchers working hand-in-hand with the military in a research project. According to A.J., "We have only scraped the tip of the iceberg."

There is much more to address about Varginha—then and now. When I went there to do my initial investigation of the case, I was literally treated like royalty. I will never forget my departure from the small aircraft that carried our group from San Paulo to Varginha. The airport "terminal" was not much of a building, but fit perfectly with the small-town ambience. Walking down the ramped stairs of the aircraft was quite a sight with numerous people gathered on both sides of the portable staircase. I was wondering who they were there to greet (it was us). At the bottom of the stairs, a man thrust out his hand and proceeded to give me a vigorous handshake. He spoke relatively good English and introduced me to another gentleman who was wearing a business suit and tie—the Mayor of Varginha! This was an unbelievable experience that I shall never forget!

Let's shift gears to the present. One of the major changes was that the magnificent new UFO Museum was never built. No other significant signs of the 1996 event were obvious. The curio shops carried no more carvings of the diminutive, brown extraterrestrials with the three protuberances on the tops of their heads. A.J. related to me that the residents of the community seemed unhappy—as if something they considered important had been taken away from them. Other details I recently learned will justify their feelings. If you will recall from my previous book on this case, I spent considerable time talking about one of the chief investigators, Ubirajara Rodriguez. (I still retain the footage of Ubirajara's office, which contained row after row of filing cabinets holding information on the Varginha case.) Ubirajara was and still is a very prominent labor defense attorney who works for some of the largest industries in the area. He is respected all over Brazil—both as an attorney and a thorough UFO Researcher.

Breaking news at the time of this writing: I have just learned that one of the two major researchers on this case has publicly denied that these events ever happened—and that person is none other than Ubirajara Rodriguez. He has also stated that he never considered the entities involved in the events to be extraterrestrial beings, but rather some strange animals. This is the very same person who, initially, wrote a book in Portuguese describing the case in extreme detail.

To make matters even more mysterious, the second investigator who was on the scene as events unfolded, the very well-respected Brazilian UFO investigator, Vitorio Pacaccini, has *vanished without a trace*. He was active in his investigations long after 1996, but has now completely disappeared.

Let us also remember that this is not Roswell of the 1947 era. These townsfolk are *angry* because of the cover-up. It is amidst this climate that James Fox, well-known producer of the UFO documentary *"Beyond the Blue,"* has recently spent a day in Varginha with a camera crew and is coming out with a remake of his original film, to be entitled, *"I Know What I Saw."*

This cover-up has angered the inhabitants of this small community and now they are rebelling. Some have not yet come forward but, as the feverish anger erupts, many have stated they will come and present their truthful testimonies and reclaim their town as the "Brazilian Roswell." One should realize that there are large numbers of both civilian *and* military witnesses that have been quiet about the matter until up to now. This author believes the adage, "Where there is smoke, there is fire," and this is just the beginning of something new and dramatic which will occur in the small community of Varginha.

CHAPTER 7

DISCUSSION

Has the world been affected by the subject of the paranormal and more specifically, Ufology? Has this subject and all its ramifications had an effect on government, world economics, academic science, wars, and our commonly-used technological devices? These are the issues addressed in this chapter.

In today's world most of the population receives its information in ways drastically different than in the 1940's up through the latter part of the 20th century. To complicate matters, not every country has information delivered in the same format. Print news still exists, but even in the poorest of third world countries, the young population derives its information from electronic media. The Internet has almost become this planet's version of the Universal Mind or consciousness itself.

The second largest subject on the Internet is the subject of Ufology. Besides the Internet, let's couple Ufology with other commonly-used forms of technology today such as the digital camera, digital video recording devices, cell phones, laptop computers and a seemingly endless array of other devices that come onto the market on a regular basis, and you have a world of recording and information dissemination such as has never been seen in the past. Countries that try to suppress UFO information through the mainstream media are slowly losing the battle to technology. The United States, the home of the Roswell UFO Crash in 1947, has suppressed information for sixty-five years and continues to do so to this day. The hidden powers that we know exist obviously want to maintain the status quo and have the methodology to exert monetary or military pressure throughout the globe. They are also privy to higher levels of information than the general public about UFOs and are still able to maintain a secure hold on it and keep it under wraps.

The vast amount of information on the subject is in the hands of governments or wealthy, private companies that are globally-based. They have used this knowledge and technology to create a hidden science that some believe may be at least 100 years further ahead of our own academic scientists. It may sound pessimistic, but if you are sitting around waiting for some country or government to let this information out of the bag, then you are going to be waiting for a very long time.

I have been asked many times if I think disclosure will ever occur. You make think my answer is NO, but the contrary is true. I do believe disclosure will occur, but not formally by any one single entity. Disclosure already occurs every day, with each photo, with each encounter, with a commercial airline crew, with each aging astronaut and cosmonaut, with each amateur astronomer, and with each scientist that breaks the bonds with his monetary master. It is happening and continues to happen—for those who pay attention.

As a UFO researcher in this field, I usually run about 5000 e-mails behind per day. This is information coming in from all over the world, not just on alien abductions or implants, but also on every phase of the subject and its related areas. Not every report is based on reality or solid science but, all in all, there are enough individuals out there who are interested enough in the subject to disseminate good information. It is up to each reader to do their own research and determine on a personal basis which stories are valid, hold their interest, or which photos are genuine. For example, I was just sent an article that has been widely viewed on the Internet by thousands of people. I forwarded this article to our research division at A&S Research to see if the subject has scientific merit. A&S Research has no opinion of this article at this time and formally denies any scientific claim as to its authenticity. But it was worth looking at. The article appears as follows:

Otis T. Carr & His Flying Machine
KeelyNet 12/23/01

Over the years, the name of Otis T. Carr comes up with regard to a flying machine he was said to have built and tested as well as a power source which he dubbed the U-tron. Scant details are available as seems to always be the case with such reports. However, in Carr's case, he did receive a US Patent 2,912,244 for an "Amusement Device."

On reading the patent and examining the diagrams it seems to be a Trojan Horse method of recording his discovery, much like the means afforded to Keely to record some of his discoveries in the form of an "occult novel" back in the late 19th century.

A search of the Net for Otis Carr and the U-tron finds several sources selling information about his work and this online document from France:

CARR, Otis T.

Otis became acquainted with the reclusive Nikola Tesla at the hotel where he resided. Tesla loved to feed the pigeons in New York's Central Park and one day instructed Otis, then studying art and working in the hotel, to buy two kilos of unsalted peanuts as pigeon food for him.

Over a period of three years, after each delivery of peanuts, Otis had the opportunity to converse with Tesla on his various discoveries. From that knowledge and his own insights, it is claimed he went on to invent a free energy generator (U-tron) and a flying device.

In 1947, Carr had finished his research on a flying vehicle (resembling the levitation disc of John Searle) and tried to interest various governmental and university agencies, all in vain because they were more interested in atomic fission.

Carr decided to direct his invention towards the educational and entertainment aspects of his work. Carr was granted a US Patent 2.912.244, for a toy apparatus which very accurately reflects the proportions and the design of his anti-gravitational flying vehicle.

The principle of operation stated by Carr was that "any vehicle accelerated towards an axis compared to its inertial mass of attraction becomes immediately activated by the energy of space and acts like an independent force."

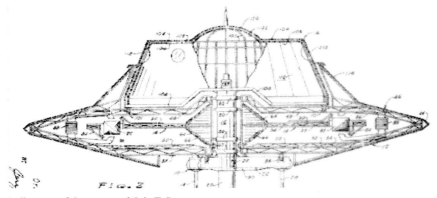

A diagram of the patent of Otis T Carr

Carr would have taken as a starting point the end of the English patent no. 300.311 of T.T. Brown by transforming the cylindrical engine into a biconic engine designed to create a non-uniform field around the ship, thanks to its design combined with its rotation.

He placed in the center another larger nipple which could be used to produce local energy and also to create another non-uniform field in the center. Although confirmed, the assertions of Carr do not explain the exchange and the extreme polarization which occurs between the iron disc with segments and the electromagnets of the circumference which give a cancellation of the gravitational field and allow an anti-gravitational flight.

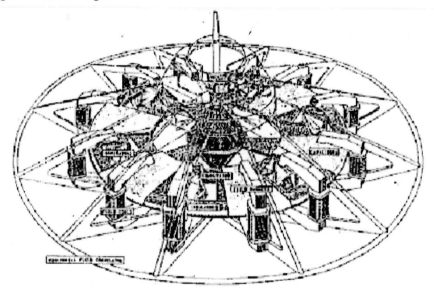

Flying Disc of Otis T. Carr
Note: 12 U-tron electromagnets in periphery

In the disc of Carr is a key component, the U-tron, which consisted of metal condensers in rotation whose form is unusual, square in a plan and round in another plan with 90°. This U-tron has the same function —rotary inductive component at high speed—with the steel plate segments of Searle.

This use of a condensing component (capacitor) in planetary rotation provided a measurable storage capacity of load at high circumferential speeds.

In the disc of Searle as in that of Carr, the accumulated load coming from the element in rotation is discharged into electromagnets on the circumference of the disc. On the disc of Carr, the spin zone contains uniform reserves of condensers which produce oscillations (pulses) for the loads as received by the magnets at the edges.

With the addition of the U-tron, Carr appears to have improved the basic anti-gravitational technology of Searle, giving them both a reinforced credibility.

The following was posted on the KeelyNet BBS as CARR1.ASC on January 31, 1991. The article is from FATE magazine, May 1958, page 17, in the regular column, *I See by the Papers*, where the following article was found.

Gravity Machine?

The following summary was sent from a Ship's Paper of October 30, 1957, after being copied from a CW News broadcast while at sea. It is unusual because no other report of this announcement reached us. It certainly is sensational – if true.

Baltimore, Md., October 29 – A group of inventors claimed Monday they have been able to utilize gravity in circular motion machines capable of powering everything from hearing aids to space cruisers.

Otis T. Carr, president of OTC Enterprises, Inc., detailed his claims in an interview and demonstration of a crude model of a circular motion machine which he said is the principle of a "free energy circular foil" space craft he can build, if someone puts up the money.

He said the machine can be adapted to devices of any size to produce continuous power absolutely free of dissipation.

Its immediate application, Carr said, would be in a space craft – which would be able to fly among the planets in controlled flight. It could land or take off as desired on the earth, the moon or any planet in the earth's solar system, he said.

Carr and his associates said their claims are based on the most simple practical applications of natural laws and discoveries in science and mathematics. They have no formal education in science or engineering.

He said the same "free energy" which causes the earth to rotate on its axis and orbit around the sun will turn a machine he described as two cones joined at their circular bases.

When the rotation of such a machine reaches a certain velocity relative to the earth's orbital velocity, Carr said, it will take off.

Carr said the core of his space ship would be a huge battery which would spin at the velocity of the external craft and which would be recharged, he said, by its motion. Carr declared such a battery, built to any size, could be designed to power the largest electric generating plant, operate an automobile, heat a house or power any conceivable machine or device.

The principle on which Carr said such circular motion machines would operate is that "any vehicle" accelerated to an axis rotation relative to its attractive inertial mass (the earth) immediately becomes activated by free space energy and acts as an independent force.

This article was originally posted on the KeelyNet BBS as CARR2. ASC on January 31, 1991. It originates from FATE magazine, August 1959, page 32.

The Saucer that Didn't Fly
by W. E. Du Soir

The serious field of UFO's and flying saucer research received a setback at Oklahoma City in late April when a highly publicized launching attempt by O.T.C. Enterprises of Baltimore, Md. resulted in failure.

Hundreds of persons had been invited to Oklahoma City by Otis T. Carr to watch him "launch a six-foot prototype model of the O.T.C. X-1, a space craft which works on "U-tron' energy." Those who were there came away disappointed. The flying saucer did not fly.

The six-foot model, well-hidden in a warehouse, never even got to the launching site, or to the amusement park called Frontier City, where it was to have been displayed beside a 45-foot version of the same X-1 which will be an amusement "ride" in the near future. The launching site was to have been a gravel pit eight miles north of Oklahoma City where Carr claimed the saucer would rise "400 to 600 feet."

Launching time was set for 3:00 PM Sunday, April 19. Two and a half hours past this time Major Wayne Aho finally announced that the launching had to be postponed because of "technical difficulties." Later it was stated by O.T.C. officials that one bearing housing was "off one-sixteenth of an inch."

Most of the major flying saucer clubs had ignored the "launching." Except for a story in early April there was no advance news on Carr's experiment in Oklahoma City newspapers. One Oklahoma City television reporter expressed the general feeling of the townspeople, "This thing will never leave the ground. And I feel that a great deal of the ballyhoo they're giving out is tied in with the ride at Frontier City. I have tried constantly to get in to see the saucer model, but they've kept it hidden."

Equally well hidden was Carr himself. He was finally located in Mercy Hospital, Room 302, by John Nebel, the famed "Long John" of WOR, New York City.

Carr did not show up at a meeting at a local church the day before the scheduled launching, but a taped speech was delivered to the 70 people who attended.

"Barring any flat tires," Carr said, "I feel that history will be made Sunday afternoon when the model of the O.T.C. X-1 is launched here."

Carr did not leave his hospital room until Monday, when he was taken to the home of Major Aho, the man in charge of publicity for the company. A scheduled "victory" dinner was pretty dismal Sunday night, where again a taped message from Carr was presented.

Monday morning at 5 AM a rumor spread that the model launching would again be attempted. At that early hour a group of a half-dozen people, including Norman Colton, Carr's chief aide, went to the warehouse. But it was three hours later before anyone else showed up and most of those who had come to see the model had given up in disgust and left.

Many believers in UFO's, although skeptical, had traveled to Oklahoma City to see what developed. All of them had left town by Monday afternoon.

One member of a Pennsylvania saucer club commented, "I don't know what's going on, but I feel they never had any intention of trying to launch the model. I could not see any plans in sight for the model and, in fact, I understand, that a Mr. Maywood Jones presented only what he called 'three-dimensional illustrations' of Carr's ideas."

Monday afternoon there was more activity in the warehouse and finally, around 4 PM, the model was "turned on" and revolved, although it showed no signs of taking off from the table on which it sat.

"We hoped to try to launch it again, but unfortunately one of the seams burst during the test," an O.T.C. official said.

"We still hope to have a launching shortly," another member of Carr's group added.

Left unanswered were such questions as:

Why was there no publicity about the launching in Oklahoma City for two weeks before the model was supposed to fly?

Why did Carr ready a taped message for delivery at the "victory dinner" a day before the scheduled launching?

How could any engineer make a mistake of one-16th of an inch in a precision part? This would be similar to a navigator landing an airplane 300 miles off course. A mistake of one-10,000th of an inch can be a major engineering calamity.

Carr was apparently undaunted by the weekend's developments. He again commented, "This indeed is an historic occasion."

Questioned about the failures of his model, he added, "The theory itself has been proven. The model will go up, if not this week, then next week or next month."

Mr. Carr also has stated that he and Major Wayne Aho will "fly to the moon in a flying saucer on December 7, 1959."

Notes about the ship:

A close-up of the left and right sides of the U-shaped electromagnet:

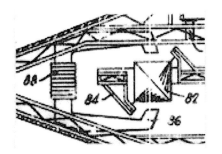 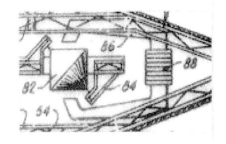

Part 82: Conically-shaped member (gyro) may be easily formed out of a suitable sheet material, such as laminated aluminum or plastic – positioned to move through the simulated electromagnets (86).

Each simulated electromagnet (86) may be formed in horseshoe shape – with tubing or, like 88, to simulate the winding of an electromagnet.

The capacitor plates from the patent:

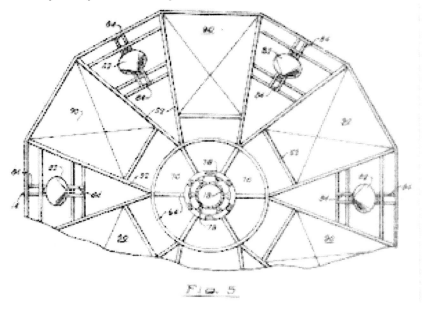

FIG. 5

Carr commented, "I further prefer to secure a plurality of plates 90 (6 plates, 6 gyros) in spaced relation around the upper face of the rotating assembly (14) to simulate capacitor plates in a space craft. The plates (90) may be formed out of any suitable material such as aluminum laminated Masonite, to provide a striking appearance."

This shows how gyroscopic motion can produce a thrust as represented by the following diagram:

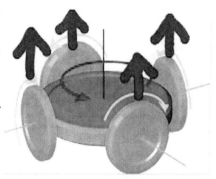

As you can see with Carr's gyros being at 45 degrees, the "beams of force" would be directed to collide at the center, above the spacecraft. Directional control would be by reducing or increasing the speed of one or more gyros on the side you wished to travel, causing the craft to "dip" into that direction.

There seems to be something else going on here, with the spinning gyros and capacitive plates being rotated through the electromagnet arms.

Now, if you look along the axis of rotation of the conical shaped members, which are gyroscopes, you will see they intersect at a 90-degree angle as shown below:

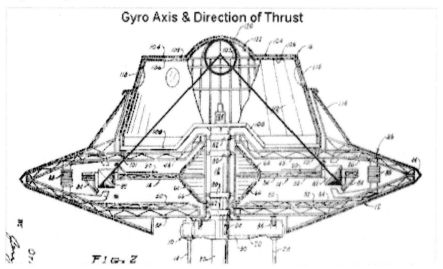

Gyro Axis & Direction of Thrust

Interestingly enough, this corresponds to two methods to alter gravity—one involves creating what Keely called an "artificial neutral center;" the other involved using a band or ring around the object, and when stimulated the ring produced weight loss (see below).

Decker's Correlations

Though there are several examples which relate, but the one I have in mind is from the claims of Dr. Daniel Fry at White Sands:

...description of two rings. When dual magnetic fields are produced in these rings, the interaction will focus between the rings. This focal point will produce an artificial neutral center towards which the mass will be attracted by virtue of resonant attraction of the natural neutral center.

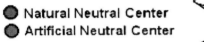

Mass natural neutral center follows the artificial neutral center

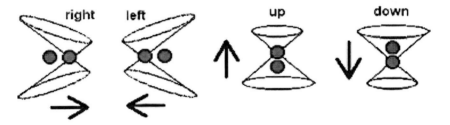

...meaning the natural neutral center of the ship will be attracted toward the artificial neutral center. It works much like a goat harnessed to a wagon.... if you hold out a stick with a carrot hanging from it, the goat will follow the stick in whatever direction you so choose.

And from Wil Wilson, an anecdote about a newspaper article with a picture describing a levitation device that was suppressed:

A cylindrical container with a series of solenoids mounted on the top of the cylinder. Each solenoid is pointed towards a central focal point. When the solenoids are pulsed at the same frequency, the combined force creates an artificial neutral center which, depending on the amplitude of energy flowing into this artificial neutral center, will cause the natural mass neutral center to be attracted to the artificial neutral center. The mass will thus be pulled in whichever direction the artificial neutral center is pointed, regardless of direction.

Another area of study is The Ship of Heaven and the aten band (more information can be found on www.Keelynet.com and its' link to The Ship of Heaven).

...This indicates the establishment of a bubble of force that protected anyone within its confines. Rather, it produced a center of gravity that absorbed aether at its own rate, rather than being chained to the planet, thus inertia and gravity were SOLELY governed by itself. Keely said you could produce this effect to such a degree that the object would float away like a bubble, independent of the mass (referring to the earth).

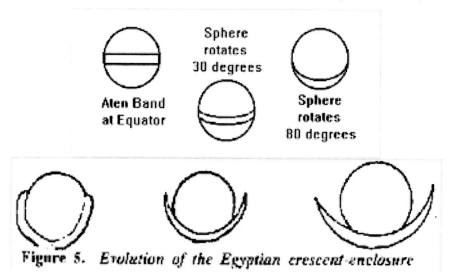

Figure 5. *Evolution of the Egyptian crescent enclosure*

Another source showing an "artificial neutral center" to deflect the aether/zpe influx that produces the illusion of gravity relates to the work of Edward Farrow. This information was sourced from KeelyNet BBS and was contributed by Bill Beaty:

"Book lost 1/6th of its weight due to action of his 'electrical condenser'...In Farrow's explanation, he said that the dynamo acted to "intensify the vertical component" of the Hertzian waves which it generated. This intensification produced buoyancy in any object to which the unit was attached. The unusual pattern of Hertzian waves fanned out in a thin horizontal plane of electromagnetic stress over a broad area.

...The condensing dynamo employed either a horizontal row or a ring consisting of a series of interrupters or breaks (gaps) for producing this field. The ring of electrical breaks extended in a horizontal line around the perimeter of the device. Power and frequency of the oscillators were not given.

The buoyant effect is similar to that produced by floating a sewing needle on water. Although the steel has a much higher density than the water below it, the surface tension permits the linkup of many surrounding water molecules in a thin film or sheet. Similarly, the dynamo lifts against the strong gravitational field by *reaction* against the weak geomagnetic field. The interaction over a very wide area between this field and the Hertzian waves produces electrical buoyancy. No U.S. patent was received on the invention."

Taking into account Carr's use of conical gryos, which would be spinning on their axis to create a force beamed at the top of the ship, they are also on rotating assembly 14 (in Fig. 5 above), which carries them through the horseshoe-shaped electromagnets. This is *highly* similar to Joe Parr and Dan Davidson's experiments with rotating pyramids.

If you have additional information you might wish to share, please contact Jerry Decker or post to the Interact discussion list. Thanks!

CHAPTER 8
FURTHER DISCUSSION

Could planetary warming and global weather change be related to the subject of UFO's? I believe the answer is yes—not that UFO's are changing our weather or warming the core of our planet, but more in the sense that these advanced and older species seem to have a way of knowing this planet's future. Perhaps part of some species' agenda is to protect the earth and its inhabitants from something that is happening in our galaxy, or at least in the solar system.

About fifteen years ago I listened to several in-depth news broadcasts predicting weather changes on all the planets in our solar system. The report was from none other than scientists with the United States Navy. They stated that our Sun was getting hotter and more active, and during the solar maximum the Sun would bombard all the planets in our solar system with high-energy particles. These would interact with each planetary body and in our case, cause a heating of the Earth's core. This, in turn, would cause a faster rotation and expansion of the molten core. This core expansion would result in more internal pressure, causing increased volcanism—which would cause a faster release of gases such as carbon dioxide, methane and other gases. Another effect would be an increased plasticity and expansion of the tectonic plates, which would begin to migrate at a faster rate, resulting in increased earthquake activity. Still another factor they predicted was the heating of the ocean floor resulting in changes in temperature of the jet stream and consequent changes in the atmospheric jet stream with concurrent weather changes. Is it not true that all this is happening? Of course there is an increase in carbon monoxide, as was recently measured. For the past 650,000 years the atmospheric carbon dioxide level was running about 300 parts per million; today it is running over 380 parts per million. This is a scientific fact.

How much of this is man-made? Some would argue that most is man-made, but scientific statistics show that more than 60% of the rise in carbon dioxide comes from volcanic eruptions. Yet, there is an increasing amount caused by man, most coming from the industrial burning of coal, gasoline and oil. Another amount comes from decaying cattle manure and other minor contributory factors. We also know that methane is another greenhouse gas and is being trapped in our atmosphere much like a glass-enclosed farming structure designed for the purpose of growing indoor crops.

The questions before us then are, "Will there be a dramatic intervention from those advanced species who seem to be our caretakers? Is there any reason to believe that they have an interest at all?" The answers seem to lie within our recent history.

A huge saucer-shaped craft was seen hovering over Chernobyl in Russia several days after the big nuclear accident there. The area was becoming increasingly saturated with higher levels of radiation and the government was evacuating more of the surrounding population each day. Suddenly the UFO appeared, and within twenty-four hours the radiation levels began to drop dramatically. Government and other sources then stated their containment attempts were successful and the impending disaster was avoided. Little attention was given to a genuine photo of the UFO hovering over the reactor.

This brings us up to recent times and the Fukushima I nuclear power plant disaster in Japan. Events were happening rapidly and an impending meltdown with nuclear detonation was predicted. As in the Chernobyl incident, the population was being evacuated at a very rapid rate. With food and water contaminated, nuclear contamination was the order of the day. Clouds of nuclear radiation were being carried aloft by winds heading for the west coast of the United States. Also, as with Chernobyl, a German news photographer took a photo of the devastated plant and in the photo was a saucer-shaped craft hovering over the Fukushima 1 plant. The photo was printed in some German newspapers and appeared for a short time on the Internet. Then suddenly, it disappeared from all news sources.

At this time I decided to do some research on my own. I looked at any and all sources I could find about the amount of nuclear radiation coming from the plant. I followed the reports for several days and, once again, all statistics on the radiation leakage vanished. Shortly afterward, it was announced on the conventional news services that some of the plant workers, wearing protective gear, were allowed to enter the plant and start the pumps again. Wasn't that a strange coincidence?

Nuclear accidents are not the only places where UFO's are seen. There are many uranium mining areas with large amounts of UFO activity—one of these being the mountainous regions of Pakistan. A few years ago nuclear missile sites were also areas of UFO activity. Some even experienced direct intervention in the United States and Soviet Union and were witnessed by U.S. and Soviet military units stationed at these bases. Today, some claim this was the primary reason for the end of the Cold War between the U.S. and the Soviets.

How soon we forget! It was September 12, 2003 and Mt. Popocatapetl in Mexico was about to explode with such a devastating force that some predicted a nuclear winter for the Earth. The Mexican government was evacuating thousands of people from the surrounding area. The earth was shuddering more each day as the cloud of ash and smoke spiraled upward into the atmosphere. Some volcanic scientists stated that this volcano has the largest magnetic field surrounding it than any other volcano on Earth. Mexico and the world's population watched the progress as if they were watching a science fiction film of the Spielberg kind. Then suddenly, out of nowhere, came a silvery disc which flew right into the heart of the volcanic beast and then came out again, speeding away at a tremendous velocity. The event was captured by a news photographer who wound up receiving the News Photo of the Year award. Within hours of the intervention by the UFO, the volcanic activity began to subside and the mountain and surrounding region once again became docile. It just so happens, as of this writing, Mt. Popocatapetl has started rumbling again and has begun to spew forth its cloud of ash and smoke.

11 Sep 2003 19:48, (12 Sep 2003, 00:48 GMT)
September 11, 2003, Mt. Popocatapetl-webcam

This leaves us with several questions: Will this major event get any coverage in the mainstream media? Will advanced beings intervene sometime again and save our collective butts? Will any government acknowledge another future intervention by advanced intelligences from elsewhere?

We shall see!

CHAPTER 9

THE TREE OF SCIENCE BEARS FRUIT

When I started my investigation with my first surgery, I thought I was helping to orchestrate a joke. Well, there was a joke involved, but the receiver of the humor was someone other than me. This was not what I had anticipated!

To date, we have performed sixteen surgeries for the removal of suspected alien implants. The sixteen victims were from different parts of the country; they did not know each other and did not have the same blood type. Also, they were not morphologically similar, had different occupations, and different family backgrounds. Some had conscious memories of their experiences while others remembered only small slices of what had happened to them.

Yet, our testing showed they all had traits which were universally found in abductees by some of our world-acclaimed researchers such as the late Budd Hopkins, the late John Mack, Yvonne Smith, John Carpenter and David Jacobs. Some of these include skin marks such as the "scoop mark," geometric patterns of dots and strange lines, ultraviolet fluorescence markings, and bruises that occurred suddenly during the night. In addition, many had reactions to electronic equipment which caused such actions as the involuntarily shutting of lights or television sets; the changing of channels without touching the TV or remote control; turning street lights on and off; involuntarily destroying FAX machines, recording devices or electronic cameras—both still and video devices. Some were not able to wear metallic jewelry because it would discolor their skin, or their metallic baubles would corrode and eventually be destroyed. Many could not wear watches of any type, as the batteries would suddenly go dead or the mechanical watches would not keep the correct time. Others had dietary habits which in general would be found in abductees. Most would eat any salty foods before anything else that they knew were much healthier choices, or they began to add salt in large quantities to other food items. Many became vegetarians. Numerous abductees developed allergies to foods such as milk, eggs, gluten, and products containing citric acid such as tomatoes or oranges. A great number of those who had conscious recollections of a specific abduction event would report being extremely thirsty when they woke in the A.M. or when they were physically returned to the place from which they were taken. In many of these cases they reported their hair and fingernails began growing at an accelerated rate. Some reported going to sleep in attire they were used to, yet when they woke in the morning, they were wearing clothing that was not theirs. Others reported taking a shower before retiring and then awoke to find the bottoms of their feet dirty, or dirt on the bed sheets. One of the questions I have been asked by abductees most frequently is, "WHY

ME?" This also happens to be a question I have been asked from audiences over the world. I have seriously thought about an answer that would at least have some grounding in scientific reason. Being a resident of the Los Angeles-Hollywood area for most of my life, I have been surrounded by people in the entertainment industry. To my surprise, there is almost an epidemic of alien abductions being performed in this area. Many of these individuals are extremely well-known movies stars. Some are artists, musicians, camera operators, lighting technicians, screenwriters, directors, producers and hundreds of others employed in the entertainment industry. I was forced to ask myself, why is this happening? The answer came swiftly: THESE PEOPLE ARE ALL RIGHT-BRAINED CREATIVE TYPES. Then I began to look at the abduction situation in general and was not surprised to find this common denominator. To make it very simple, the one single thing I have found that abductees have in common, worldwide, is their creativity. Our abductors are looking for human creative thinkers. They don't seem to be interested in mathematicians, bricklayers, bankers, accountants, or generational histories of left-brained individuals.

Let me stress this very important point: most abduction cases appear to go through family generations. I have seen cases with implants where I have been able to find a similar object in a grandparent or even in a deceased great-grandparent where x-rays were available prior to that individual's death.

Why are implants necessary to an advanced species? I believe we now hold the key that will unlock this question, and the answer is very simple. If we study the alien abduction scenario on a world scale, we find that over 90% involve the taking of ova and sperm. That simply means there is something genetic happening. I believe that the human race is having its DNA altered and we are becoming a race of new beings. There was a name given to this "new being" by John White, book agent, promoter and organizer of multiple large UFO conferences in the state of Connecticut some years ago. The term was: **HOMO NOETICUS.** Some of us are still Homo Sapiens, but the newer generations are Homo Noeticus. Proof of this is simple to find. Listen and look closely at what your children and grandchildren are saying or doing. You will be shocked. Look at the major changes going on all over the globe, such as in the Middle East. The uprisings are mainly instigated by the younger generation using modern advanced electronic technology which they grew up knowing how to use. The Internet itself is the consciousness of the young masses and they are putting it to use. If there is a salvation to this world's state of affairs, it will be up to the younger generation—**HOMO NOETICUS.**

Why are implanted objects needed? If you will recall, there was major news during the early years of the United States manned space program which revealed that some of the astronauts were required to swallow implants. Some didn't like it, but they were required to do it regardless of their reticence to do so. Why was this necessary? The only way for Mission Control to gain information on the physiological state of the astronaut was for him to swallow devices that would broadcast the specific data they were looking for in Mission Control. Through this means they could determine exactly what was going on in the bodies of humans in space. We can logically relate this to an advanced race of beings that might be doing DNA manipulations on humans. Perhaps there is important data they need in order to determine the progress of their genetic experiments without reacquiring the human that was manipulated. Hence, they implant tiny devices that will relay this information to them while they are still in close proximity to our planet.

Remember, in the early chapters of this book I spoke about some of the devices that we have removed which were emitting radio signals in our electromagnetic spectrum. Also remember that in one instance I was able to determine that this signal was from a fixed or mobile device operating on a deep space frequency. We then wondered why an advanced civilization would use a primitive radio wave to carry messages. From a theoretic standpoint, in the sci-fi classic *Star Trek*, the Federation and other advanced races transmitted information over long expanses of space using sub-space frequencies because they worked the best. These types of frequencies could be useful in real life for a number of reasons. Perhaps these entities are using scalar wave frequencies and what we are detecting here is only a harmonic of the original scalar frequency. And we must consider the possibility that knowledge is being shared with some group of humans, perhaps as part of the Black Budget program. It is also possible that more advanced races have technologies that are ahead of humans and other technologies that are not—so must be replaced with something we have. A "dumbing-down" of their more advanced technologies would be necessary in order for them to work on a joint project. We must all realize, in all truth, we do *not* have the answers and are really only scraping the surface of this complex subject.

I cannot over emphasize the following:

1. As humans today, we do not know the truth of where we came from.

2. Our science is primitive and we still live in the primitive science of yesteryear.

3. We have been lied to over the centuries by the masters of our race.

4. Our culture is still one of slavery: we are all slaves to a higher master whether it is our boss at work, the almighty dollar, or dozens of so-called creators of man that we call our Gods.

5. We have even been kept in the dark about our own history which seems to change at the whim of the historian. Many of our events were chronicled by humans who were never there.

6. We run from the truth as fast as we can because it might interfere with our routine activities. We are purposely subverted by emphases on sports, the movies, and of all things, the intake of food.

7. We are rapidly becoming slaves to electronic technology with a myriad of new gadgets coming onto the market each week.

8. Our children are losing their ability to interact on a personal level because they rely on electronic devices to communicate.

9. Our imagination has been torn from our minds and replaced with repetitive auditory and visual stimuli.

When will this change? Are we a race in crisis? Do we have any control over the future when our past is basically unknown?

My answer is not a prediction of gloom and doom. The world is in the hands of the new humans, the younger generation, *Homo Noeticus*. It will be up to them, working with advanced races, to salvage this planet and attempt to restore the earth to a paradise, as similarly mentioned in the Bible's book of Genesis.

CHAPTER 10

WHO KNOWS ABOUT THE IMPLANT RESEARCH?

It may come as a great surprise to you, as it did to me, to learn that my work has been discussed in the Capitol of the United States, in the White House, and shown to President Obama. My dear friend John Greenewald pioneered the web site called The Black Vault. When he was a very young man, he began collecting documents under the Freedom of Information Act. He started with several documents and now the website boasts over one hundred thousand. Yes, some are severely redacted, but others contain thousands of pages which are not redacted at all.

To my great surprise, John sent me a document he just received not long ago which pertains to me. I opened it and was shocked at what I saw! I am reprinting the document here, just as it was attained from John Greenewald through The Black Vault. Please read with interest as it is all the *truth*. All the "*i's*" are dotted and the *t's* are crossed." It is one hundred percent accurate. Following the introductory letter, directly below, it consists of a series of inter-office emails within the White House. This e-mail correspondence was sent to The Black Vault Website at the request of John Greenewald along with the official OSTP Report. They are inter-White House communications and most of the peculiar lettering and symbols have not been redacted. This is the first time that the public can be exposed to the complex workings of the inter White House staff.

EXECUTIVE OFFICE OF THE PRESIDENT
OFFICE OF SCIENCE AND TECHNOLOGY POLICY
WASHINGTON, D.C. 20502

March 22, 2011
Mr. John Greenewald, Jr.
RE: OSTP FOIA **11-14**

Dear Mr. Greenewald:

On March 8, 2011, you sent the Office of Science and Technology Policy (OSTP) a request under the Freedom of Information Act, 5 U.S.C. § 552, requesting all documents pertaining to the UFO Phenomenon. OSTP received your request on March 8, 2011. Your requested documents are enclosed.

If you consider this to be an inappropriate denial of your request, you may appeal in writing within 30 days of receipt of this letter to General Counsel Rachael Leonard at ostpfoia@ostp.eop.gov or via fax at (202) 395-1224. The email should be clearly marked "Freedom of Information Act Appeal."

Sincerely,

Diana Zunker

Wells, Damon R.
From: Noveck, Beth S.
Sent: Monday, June 22, 2009 9:23 PM
To: Samanta Roy, Robie I.; Kohlenberger, James C.
Cc: Wells, Damon R.
Subject: RE: UFO strategy

Have you read the article and re-read Jim's email?;-)

From: Samanta Roy, Robie I.
Sent: Monday, June 22, 2009 9:18 PM
To: Noveck, Beth S.; Kohlenberger, James C.
Cc: Wells, Damon R.
Subject: RE: UFO strategy

Beth,

Check out this website: http://www.nasa.gov/offices/hsf/contact.us/index.html

This is the mechanism that NASA is using to solicit inputs from the public for the human spaceflight review that OSTP chartered (the Augustine Committee). This website has a few apps ranging from ingesting documents to submitting (and voting on) questions for the Augustine Committee (that are then subsequently answered on-line). Last week alone there were 52,000 page views and about 1,000 followers on Twitter.

Regarding the term "space policy", that's a big field covering a broad spectrum of civil, commercial, military, as well as international issues. For NASA human spaceflight issues, we could definitely link up w/ the NASA site (and/or get them to use some of the OSTP website apps), but for other issues, the NSC has kicked off a classified review of Administration national space policies, so we should defi ve a chat w/ them before we decide how big is our aperature (on serious issues beyond UFOs). I'm out on duty this week, but can telecon in early in the morning (before Bam), at noon, or in the later afternoon (or beyond). (.b}Lb)

One potential idea to get our blog started: during the campaign, the President released a space policy paper calling for a National Aeronautics and Space Council in the White House. Perhaps we can start asking for ideas on this topic... Thanks.

–Robie

From: Noveck, Beth S.
Sent: Monday, June 22,2009 8:18PM
To: Samanta Roy, Robie I.; Kohlenberger, James C.; Wells, Damon R.
Subject: RE: UFO strategy

Once you read the NYT article, the connection between our open government strategy and our space policy will become immediately clear. In fact, we'll want to draft an announcement for the blog tomorrow.

From: Samanta Roy, Robie I.
Sent: Monday, June 22, 2009 7:59PM
To: Kohlenberger, James C.; Wells, Damon R.
Cc: Noveck, Beth S.
Subject: Re: UFO strategy

Ok will check out later this evening....Beth can you forward the ideas or do we find them on the blog under some heading?

From: Kohlenberger,James C.
To: Samanta Roy, Robie I.; Wells, Damon R. Cc: Noveck, Beth S.
Sent: Mon Jun 22 19:44:37 2009
Subject: UFO strategy

R&D –

I need you guys to get together with Beth and work some of this through. See the attached story. There are some interesting options coming through the open government blog that deal with space flight. Some of the ideas might reduce the space flight gap. Can you work with Beth to evaluate these options – she has apparently identified a number of experts.

Obama, Inviting Ideas Online, Finds a Few on the Fringe*[author note: this was a link]*

Jim Kohlenberger
Chief of Staff
Office of Science and Technology Policy
Executive Office of the President
202- (b)'")

Wells, Damon R.
From: Noveck, Beth S.
Sent: Thursday, July 09, 2009 11:11 AM
To: Kohlenberger, James C.; Wells, Damon R.; Samanta Roy, Robie I.; Stebbins, Michael J.; Weiss, Rick
Cc: mwbaldwin@ gmail.com; Sturm, Robynn K.
Subject: RE: Who has the lead?

Oh, yes, this is DEFINITELY up my alley. We'll get right on it!

Sincerely yours,

Director, Intergalactic Policy.

From: Kohlenberger, James C.
Sent: Thursday, July 09, 2009 9:01 AM
To: Noveck, Beth S.; Wells, Damon R.; Samanta Roy, Robie I. Stebbins, Michael J. Weiss, Rick
Subject: Who has the lead?

I'm trying to figure out who has point on this issue? Any guidance ... health care reform is a priority for us.

Special Note: The article that appears a few paragraphs below was attached at this very spot in this email. The "Key to Obama health plan..." headline appeared here, directly under this last sentence (above). The original article first appeared two weeks previous on www.examiner. com and was seemingly sourced from this site. It was apparently not used as a government document until after it was sourced from the Internet. It appeared in this email, then later released through the Freedom of Information Act to The Black Vault, which collects these sorts of things. Government documents, as many of us know, do not have copyrights. This article did originally have a copyright, but when released to The Black Vault, the copyright notice no longer appeared. It was only due to the further research of my publisher and myself that we found who had originally written this piece: long-time investigative journalist and UFO researcher Larry Lowe. This document, as of this writing, can still be viewed at www.examiner.com, with copyright notice still intact. FOI documents do not have copyrights, and this one was seemingly stripped of its protected status by appearing in this way, through these channels. Because of this, we at least wanted to credit the original source. A FOI request will apparently get you the copyright free version. We are aware of a similar case that went to court whereby the "no longer applicable copyright" was upheld. In trying to sort this out and not get sued for copyright infringement, I contacted the author of this piece. He clarified the situation in an email, and suggested the following explanation.

"In June of 2009 I, Larry Lowe, investigative journalist, wrote a piece for Examiner.com detailing the hypocrisy of the Obama administration in pushing hard for health-care reform while ignoring the potential medical benefits implied by the work of Dr. Leir in examining enigmatic implants, which, among other unusual characteristics, did not provoke any response from the human immune system. The ability for devices or tissues to be implanted without any signs of rejection would be a leap in medical treatment capability. The piece also discussed the process of marginalizing feedback from the public the administration did not want to hear."

"The article, reprinted below in its entirety with my permission, including the copyright and permissions block, had an impact unbeknownst to the author. On March 22, 2011, in response to a FOIA request by John Greenewald to the Office of Science and Technology Policy, the entire text of my article appeared in an email from James C. Kohlenberger, then Chief of Staff of the OSTP in the body of an email from him to his staff, asking "Who has point on this issue?" indicating a level of political concern at the least, over the issues raised in the piece. Kohlenberger's email left out two small but vital pieces of information: the author's name and the copyright and reprint block at the conclusion of the article. By doing so, Kohlenberger violated the copyright notice he had excised from the copy. This led some to interpret the FOIA document to indicate the Chief of Staff of the OSTP had written the piece, when in fact he had copied it from the webpage at Examiner. com word-for-word, without attribution or permission... If plagurism is a form of compliment, it's then highly gratifying that my discussion of the subject was copied word-for-word in a White House email, only to emerge years later via FOIA."

"So, while the White House officially denies any knowledge of the UFO issue, the Chief of Staff of the Office of Science and Technology Policy considered the potential implications of possible medical advances that could result from free and open examination of implant cases significant enough to alert his staff to the issue. Kohlenberger went on to say in his email 'health care reform is a priority with us.' Transparency is apparently much less of a priority."

Key to Obama health plan blocked by UFO truth embargo?
June 25, 3:10 PM

Photo of President Obama at National Press Club. (Note: The original photo in article was from the annual meeting of the American Medical Association, Monday, June 15, 2009 in Chicago, however, poor newspaper quality prevented its usage.)

Tiny fragments of a device of apparent extraterrestrial origin may hold an invaluable secret to advancing health care in the US, but open discussion of that fact may be subjected to information policing by the Obama administration.

The sad irony of the situation is that a president who campaigned and won on a platform of truth and openness and who has just presented a comprehensive plan on ABC in prime time to reform the nation's medical care may unwittingly be preventing discussion that could lead to a breakthrough which would revolutionize health care.

Complex miniature devices of unknown and apparent extraterrestrial origin have exhibited a remarkable ability to be assimilated into the human body as foreign objects without inflammation or rejection by the human immune system. The mysterious coating on these objects, if fully understood, could revolutionize organ transplant and other operations that require pins, screws and plates to be implanted in the human body.

The New York Times, however, is reporting that while the Obama administration has asked the public for new ideas with the unveiling of an open-government website, the administration is considering steps to curtail free speech and discussion of the UFO issue on the prejudiced supposition that UFO subjects are somehow 'fringe' – the modern term for heresy, which was considered a crime against the church when Galileo published his findings in opposition to the conventional wisdom of his time.

The White House tried to screen out some of the more unusual comments in the second phase of the process. Ms. Noveck summarized the most significant ideas, then invited comments on them at blog.ostp.gov. Visitors could flag off-topic comments, which were then shunted to a separate part of the site. That reduced the birth certificate and U.F.O. comments to a relative trickle. —The New York Times

The argument goes – from guys like Clay Shirky, who ought to know better – that important public debate should not be 'hijacked' by a vocal minority. This view is more likely ill informed than disingenuous, but in either case it flies in the face of the facts. The UFO issue can hardly be painted as the interest of a 'fringe' minority. Polls show that 50+% of the American people believe some UFOs are extra-terrestrial spacecraft and 85+% believe their government is not telling the truth of the matter.

Suppressing free speech on matters such as physical evidence of UFOs uncomfortable to the psyche of a select few who would monitor the content of the public debate is an insidious form of censorship.

The spirit of the first amendment aside, what is at stake is a health care breakthrough that would save millions of dollars and thousands of lives – if research is funded and accomplished to understand why the mysterious devices exhibit an unusual ability to assimilate themselves *inside* the human body with no reaction from the immune system.

The fragments in question are the remnants of an enigmatic device removed from the toe of a subject which was apparently some kind of tagging and tracking unit. Veteran implant-removal specialist Dr. Roger Leir made the announcement at a press conference held by the Paradigm Research Group immediately after the 2009 X Conference in Washington, D.C. Leir's presentation of the biological nature of the mysterious implants was complemented by the results of a material analysis conducted by Dr Alex Moser, Ph.D.

The implications of the material's composition and apparent function are paradigm shifting. At the time, the implant analysis report was overshadowed by Apollo 14 Astronaut Edgar Mitchell's statement of belief that ET has been here for some time and that the elements of the government who actually know this have been keeping that information highly compartmentalized. Since then, however, a highly credentialed physicist has spoken out on the issue, a nationwide radio host has investigated the report and the story has begun to gain legs.

For just-the-facts UFO investigators, Leir and Moser offered the compelling news of the 2009 X-conference press conference: hard evidence that can be analyzed and results that point toward non-human origin of the device.

Due to a number of unusual characteristics documented in the analysis, there are only two conclusions that can be arrived at as to the origin of the implants:

Either an unknown agency of extraterrestrial nature is manufacturing highly sophisticated biomechanical devices and implanting them inside human bodies,

Or some terrestrial agency with enormous resources has achieved nanotechnology fabrication capability beyond currently known limits and is doing the same thing.

Either possibility is unsettling.

That the first possibility is a credible concern alone is enough to be front-page news. The mainstream media truth embargo about UFO/ET evidence or the amount of science required to present the story may account for the fact it wasn't.

Here is what we know:

Background

In 2003 a sophisticated use of the Roper Organization by Bigelow resulted in a solid estimate that of about 2% of the population has likely had an experience of the nature termed abduction. In order to avoid any kind of data skewing it was necessary not to telegraph the nature of the inquiry. Certain questions designed to reveal characteristics pointing to abduction were inserted into in three separate Limobus surveys conducted by Roper. A false positive trigger question was included to identify those who could not have been abducted but were giving positive answers.

Successful abductions include selective amnesia of the event and thus accounts of abduction often need to be retrieved by hypnotic regression. A large percentage likely goes unreported.

Roper's representative American sample of about 6000 adults (with a sampling error of 1.4 percent!) showed that one out of every 50 people met the abductee profile. This

figure suggests that about 33,000,000 individuals had been abducted in America. A closer look at these specific profiles showed that these people were not "average" at all. —Report on the Roper Analysis Data

A characteristic of the accounts is the implantation of a small device at various locations on the body. The presumed reason for this would be subject tracking and data gathering – a process humans have developed while gathering information about the other species inhabiting our planet. Once you consider an advanced intelligence of some form operating in our biosphere, the notion of small implants for data capture and tracking is a trivial leap.

Dr. Roger Leir is an acknowledged pioneer in the field of implant detection and removal. Since 1995, he has conducted 15 surgeries to remove enigmatic objects from subjects who reported abduction experience by non-human entities. These surgeries removed miniscule metallic devices detected by X-Ray. The objects themselves constitute some of the rare hard evidence of extraterrestrial origin – and thus technology and thus intelligence – available in the wake of tens of thousands of UFO sightings annually worldwide.

The most recent and significant of the implant removal cases is the object removed by Leir last year and presented at the conference.

The Devices

While in the body, a device is identified on X-Ray, located precisely with a CAT scan. It shows up as a metallic object where one should not be. There is no apparent entry point, which implies that the object was either inserted with

Microscopic View of Implant Device

an advanced form of medical implant technique or somehow grew in place. The latter is highly unlikely given the complex nature of the device.

The devices exhibit a measured magnetic field of 5 mGauss. A refrigerator door magnet, by comparison is 15 mGauss. There are no natural objects in the human body we know of that have magnetic fields anywhere near this strong. There are some magnetic fields associated with the operation of the nervous system, but these are extremely weak (on the order of micro Gauss, or less).The earth's magnetic field is approximately 300 mGauss, in most areas, so this is about 1/30th of the earth's magnetic field strength.

Alex Moser speculates that the magnetic field of the device may have a function in powering it, via zero-point energy, or in helping to generate the radio signals it was emitting before removal from the subject.

DR. LEIR AT NATIONAL PRESS CONFERENCE IN WASHINGTON D.C. SPEAKING ABOUT THE IMPLANT DEVICES. DR. EDGAR MITCHELL WAS ALSO PRESENT AND SPOKE ABOUT HIS KNOWLEDGE OF THE EXTRATERRESTRIAL PRESENCE ON EARTH.

The devices emit RF transmissions on several frequencies, ranging from the extremely low frequency of 9 cycles per second, which corresponds to a naturally occurring human brain wave frequency up to the gigahertz range.

The 9 Hz ELF frequency observed is said to have applications in mind control. The 17 MHz signal is in an aeronautical mobile (aircraft communication) band, and the 20 GHz frequency is in a satellite communication band.

While Moser thinks that the implant may have had mind influencing/control as one of its functions, he also considers it possible that aliens use our communication bands for some of their own (heavily encrypted) messages, perhaps to be able to more easily listen in on human communications.

No recordings of the RF emissions has been made and no attempt made to do signal analysis of the emissions, so what information may be being broadcast, if any, is unknown at this time.

As for power output, Moser reports that *"we do not have quantitative data on the signal strength, but an educated guess as to the power output would be 10-100 mWatts. The device was apparently transmitting continuously, unlike an RFID chip, which only transmits a very low-power RF burst, when irradiated by a nearby transmitter. Monitoring of the signals did not take place for a sufficiently long time interval to determine whether the transmissions were influenced by the subject. The signals ceased after removal from the subject. In this case, the cessation of transmission may have been due to the implant breaking into pieces on removal, but the RF signals also ceased in previous implants which were removed intact."*

The power source for these embedded devices is unknown. One assumption is that power for the device is somehow biochemically extracted from the surrounding body. A second theory from Dr. Koontz contends that the low frequency emanations from the device may be the signature of a scalar energy extraction system – which could lead to an understanding of propulsion systems required to travel to the stars.

The interface between the unusual object and the human body is a thin later of material which connects the object with the surrounding tissue in such a way that the foreign body becomes part of the human body. There seems to be a coating on the outer edge of the object that facilitates this process. The potential for this coating to remove the issue of inflammation and rejection of implanted material in the human body is significant. All medical procedure that involves stitching together the human body with pins, screws, stints and plates would benefit from this coating.

No known incident explains the presence of the object in the subject's body. We don't know how it got there, we don't know why it wasn't rejected, we don't know what powers it and we don't know why it is emitting RF transmission on range specific frequencies.

The Material

Once removed, analysis of the object fragment indicates a level of material fabrication beyond the limits of publicly announced human capability. It also demonstrates material components not consistent with materials found on earth.

Included in the complex structure are carbon nanotubes, small molecular constructions of carbon atoms, which terrestrial science is just beginning to understand and utilize.

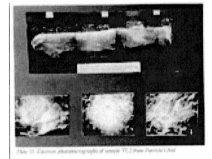

One of the most significant indicators of extraterrestrial origin is the variance of isotopic ratios of the composite elements from normal terrestrial elements. If you take an element such as Nickel or Silver, there are several different isotopes of that element available in nature. The isotope of an element

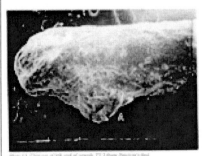

refers to the specific arrangement of the electrons orbiting the nucleus, and a moderately heavy element like a metal may have the same number of electrons arranged in different numbers in the bands surrounding the nucleus.

The ratio of common isotopes of an element is fixed and common to the element on earth, however specimens recovered from space exhibit different ratios of the isotopes than terrestrial atoms.

The extracted implant material exhibits isotopic ratios at variance with that of terrestrial elements. If this measurement is correct, the implication is that the only way this material could have been manufactured is if the fabrication facility had access to a supply of extra-terrestrial materials or the technology to produce non-terrestrial isotopic ratio metal – an unlikely possibility.

When researching this kind of UFO evidence, however, one must proceed carefully. In an email, Moser expressed a concern that one of the measured isotopic ratio anomalies may be instrumentation or measurement error.

Moser's caveat is that *"as stated in the report, the analysis lab could not give adequate explanation for the Ni isotopic anomaly. However, I am fairly certain the error is derived as described in the report (beam broadening as a result of high nickel loading within the analyzer)."*

This does not explain all of the isotopic ratio anomalies, however, and further research is needed.

Prior Isotopic Ratio Anomalies

The 2009 Leir implant removal is the first time such material has been detected in an obviously fabricated operational device inside a human being, although there have been reports of isotopic anomalies in material alleged to have been recovered from a crashed ET spacecraft.

A 1997 report in CNI News discusses in detail the major prior instance of isotopic anomalies in alleged extraterrestrial material, which has not been followed up with adequate verification research:

On the morning of July 4, 1997, in an auditorium in Roswell, New Mexico, hundreds of news reporters and other interested onlookers came together for what was billed as a press conference on the scientific testing of an object said to have been recovered from the crash of a UFO near Roswell in 1947.

The main speaker, Dr. Russell Vernon Clark, a chemist from the University of California at San Diego, delivered prepared comments and then immediately left the auditorium, frustrating many journalists who wanted to ask him questions. Even so, Vernon Clark's announced findings undoubtedly represented the biggest surprise of the week-long festival called Roswell UFO Encounter 97. —CNI News, 1997

In an obvious error in the 1997 results was in the isotopic ratio of Germanium, which has a half-life of a couple of days. This result precluded the sample being old enough to have had an origin in 1947 and was obviously a measurement error of some sort.

Moser explains that *"I reviewed a small portion of [Vernon Clark's] data set and was concerned of the claims because I felt the analysis method used was inadequate and the data set showed behavior that indicated inadequate sample signal."*

Concerns about the error rate of the measurement on the isotopic ratio evidence make Moser justifiably cautious, but the presence of carbon nanotubes, a magnetic field, and unusual radio frequency (RF) emissions from the embedded device, coupled with the magical ability to reside in the body without provoking a rejection response from the immune system, not to mention the inexplicable origin of the devices make this evidence worthy of considerable scrutiny.

Separate from the isotopic ratios, the issue of the ratio of various *rare* elements indicates a possible extraterrestrial-origin of at the least, the raw materials from which the device was fabricated.

Moser states that *"The gallium, germanium, and precious metals in the metallic portion of the sample were present in ratios which were very consistent being derived from an iron-nickel meteorite.*

There is a lot of this material in our solar system, and it would be a good, naturally occurring magnetic material, which would be inexpensive to an organization with cost effective space flight capabilities. It would be a good base material for aliens to use for fabrication of a device requiring a magnetic field.

There is probably some type of low-temperature melting process involved in getting the carbon nanotube inclusions/ components inside the metal of the device, to avoid conversion of the CNTs to iron carbide, if the metal had to be melted at its normal melting point. Some type of automated nanoassembler would also almost certainly be involved in the fabrication of this type of device."

The Letter of Support

Shortly after the announcement, a letter of support came from a surprising source. Experimental nuclear physicist Dr. Robert W. Koontz, Ph.D. is an expert in carbon nanotube fabrication whose published credentials lend considerable weight to his opinion that this object may represent, among other things, a planetary security issue. That likelihood must, he says, be openly discussed and understood by the public.

In particular, I note the reported non-terrestrial isotope ratios of the putative implant, the reported emissions of electromagnetic energy and the apparent micro-structure of the possible device. This is physical evidence that has been and can be analyzed.

1 also note that the interviewed scientist seems quite clear-headed and sensible. Furthermore, the scientist has demonstrable knowledge about carbon nano-tubes and appears to indeed be the scientist he claims to be.

I see no reason whatsoever to discount what these men are saying. Indeed, quite the opposite is true: My opinion is that this matter should be taken very seriously and, eventually, should be openly addressed by both federal authorities and by the public. —Dr. Robert W. Koontz, Ph.D.

Moser's response to the Koontz open letter is positive: *"I read the letter and I agree with it in its entirety. Though I believe it is possible the situation is vastly more complex than most understand."*

The unusual step of publishing his credentials on the web and speaking out on the so called "fringe" issue is an example of the American citizen no longer willing to let compelling evidence be swept under the rug of ridicule, denial or marginalization by proclamation, such as practiced by Shirky and the New York Times. The issues associated with the increasing evidence of UFO and ET activity in our biosphere must be addressed openly. Anything less leads to a debilitating cultural schizophrenia.

The Remaining Questions

The applicable Carl Saganism in this case usually is that *"extraordinary claims require extraordinary evidence."* Dr. Roger Leir has presented extraordinary evidence. At this point the corollary is that *extraordinary evidence requires extraordinary investigation.*

A research organization wishing to remain anonymous checked a number of subjects with possible implants at a recent UFO convention, using a much more sophisticated radio frequency spectrum analyzer. Activity was detected at the same frequencies (except ELF, which the instrument was not sensitive to) plus some not observed previously. However, according to Moser, a lack of serious funding prevents *"the kind of study this evidence requires."*

Eisenhower warned that when research is a slave to government funding true intellectual curiosity would be diminished. This is clearly such a case, for the medical benefits of understanding the mystery coating alone would save countless lives. Yet the extraordinary research demanded by these enigmatic objects will never come from a military-industrial complex that seems determined not to allow academic investigation of anything implying an extraterrestrial intelligence operating in our biosphere.

The potential medical benefits aside, the implications of the origin of these devices confront the crumbling paradigm that we are somehow inexplicably alone in a vast universe that is likely teeming with life. Who made these devices? Who implanted them in the human subjects without their knowledge, much less consent?

What information is being gathered and who is benefiting from that information? How many of us are implanted and don't know it?

If it is a form of advanced intelligence, what are they trying to find out about us and how do we deal with the fact that they are? Can these data gathering devices point a way towards communication with whoever is doing this?

Should SETI be thinking about transmitting on these frequencies messages intended to get a response, in order to let the chipmakers know that we know that they are implanting us?

And if it is not some form of advanced intelligence studying us, then what secret element of our society, what agency is covertly tagging, tracking *and* studying its own citizens? Who approved the policy? Who provided the funding?

When will a member of the White House correspondents' corps finally have the temerity to ask a question about this issue?

The Whitley Strieber Interview

At least one journalist is not afraid to pursue the story on the national stage. Whitley Strieber interviewed Ph.D. Robert Koontz on his Dreamland radio program. The conversation, in light of the questions above, is compelling listening for anyone who understands what has been presented in this article.

One of the things that can be done is to launch a Twitter initiative to ask for adequate funding of research into the 2009 Leir implant remains.

The truth embargo maintained by the powers that be is failing; *the race to disclosure has begun.* Discovering and disclosing the extent to which everyday citizens are being tagged and tracked – and by whom – is a critical step.

As part of the recently proposed sweeping health care reform bill, the Obama administration should include funding to detect, extract and study these objects with an eye at the very least to reverse engineering the biological coating.

And someone should tell the Obama administration that suppressing free and open inquiry into aspects of the UFO phenomenon that may hold the key to significant advances in the human condition is *not* the way to make good on the inauguration day promise of truth and openness.

The material I have presented to you is scientific proof that advanced civilizations have been visiting the earth today, as well as in our recent and distant past, and have intervened with life on this planet for, quite likely, millions of years. The final chapter, which follows, presents in-depth information on Dr. Robert Koontz, Nuclear Physicist.

CHAPTER 11

IN FINALITY

Dr. Koontz is a high level scientist with extremely impressive credentials, some of which are documented in this chapter. Some of his work has been so highly classified and involves such advanced and sensitive research, that he is not allowed to discuss it to this day. Below is Dr. Koontz's original statement relating to his study of an implant:

Full Report On the Alleged Alien Implant Removed From an American Scientist Code-Named "John Smith"

Explanation of Galactic Distance

Some time ago I wrote an open letter regarding the scientific findings of a possible implant that was removed from a scientist who has been code-named "John Smith." (*author note: see page 143*)

However, one key finding was that the putative implant that was removed from "scientist John Smith" had isotope ratios which were very much different than those normally found in terrestrial materials.

I thus believe it is proper to say that the removed object has (or had) non-terrestrial isotope ratios. **That would be consistent with the object having been constructed in a stellar system other than ours**.

Regarding the 2.2% isotope shift of the Boron-10 isotope in the examined object, if the shift is due to a weak electron-capture process that leads to Beryllium-10, then the object could have come from a stellar system which began to evolve approximately 100 million years in advance of our solar system.

That would be $00.022 * 4.6$ Billion Years or roughly 100 million years.

It is thus conceivable that the alleged implant came from a civilization that began its stellar evolution roughly 100 million years in advance of ours.

Robert W. Koontz
Experimental Nuclear Physicist

Dr. Koontz is saying that the extreme differences in the isotopic composition of the object make it apparent that the material originated in a solar system other than our own.

Dr. Koontz then used the fact that Boron 10 can decay by electron capture to Beryllium 10, with a known rate, to try to give a relative age of the Boron in the implant material, and then used this difference in age to try to estimate how far away the source of the material was from Earth.

He used the 2.2% difference in the Boron 10 amount in the sample, relative to the Boron found on Earth, to calculate the difference in age, from the rate of this radioactive decay reaction. A 100 million year difference in the age of the solar system the implant material came from was derived from this calculation.

Dr. Koontz then multiplied this age difference (100 million years) by a physical constant called the fine structure constant, which is dimensionless, and has a value of approximately 1/137.

100,000,000 years * 1/137 = 73,000 light years

So, by this method of calculation, the material in the implant came from approximately 70,000 light-years away, possibly from the other side of our galaxy.

E-MAIL FROM DR.KOONTZ

The alleged age of the universe is 13.7 billion years. And the fine structure constant is ~ 1/(137.03). Therefore, if we multiply the age of the universe by the fine structure constant, we get 100 million years, which is the time I calculate in association with the implants.

That is a galactic super cell distance. 1/(137.03) times that is a galactic distance.

So, again, we are talking about galactic distances, not distances to stars.

Well, we would be going from relative isotope shifts to distance and I don't quite know how to do that. You could multiply 90 million years by the speed of light and get 90 million light-years and then multiply that by the fine structure constant. That would end up being 650,000 light years – which is about the size of the largest galaxy you will find. Then you could divide by a factor of 2. This would be consistent wit measuring distances in the phase-space time of electrons.

So, my guess is that we can say we are talking about galactic length scales. That's the best I can do just now.

I would say that ~300,000 light years is a good guess.

RWK

Open Letter On Scientific Evidence for Extraterrestrial Implants
by Robert W. Koontz

May 26, 2009

To Whom It May Concern:

I am a Ph.D. experimental nuclear physicist, and I was once with the US Navy's Naval Security Group. While assigned with the National Security Agency, I taught electronics related to remote intelligence gathering. My clearance is a lifetime National Security Agency Top Secret with Cryptographic Endorsement and Code-Word Access.

In the web page linked to below, I have posted news articles and background information that substantiate my credentials.

http://www.doctorkoontz.com/bio/Deep_Background/index.htm

Regarding Whitley Strieber's reports about alien implants and his recent interview of Dr. Roger Leir, and also regarding Whitley's interview of an American scientist who says he was implanted with some sort of technological device, I find the evidence very compelling.

In particular, I note the reported non-terrestrial isotope ratios of the putative implant, the reported emissions of electromagnetic energy and the apparent microstructure of the possible device. This is physical evidence that has been and can be analyzed.

I also note that the interviewed scientist seems quite clear-headed and sensible. Furthermore, the scientist has demonstrable knowledge about carbon nano-tubes and appears to indeed be the scientist he claims to be.

I see no reason whatsoever to discount what these men are saying. Indeed, quite the opposite is true: My opinion is that this matter should be taken very seriously and, eventually, should be openly addressed by both federal authorities and by the public.

However, I realize that federal authorities are unlikely to openly address this matter, and my opinion is that mainstream news media will not write even a single, unbiased, article on the subject.

Nevertheless, if it is true that extraterrestrial persons are placing implants in the bodies of US citizens and US scientists, then the matter is of a national security nature that could be more serious than the threat from al-Qaeda and North Korea.

It is possible that my comments will be met with mockery and derision in some quarters. But that does not dissuade me in the least. Let the chips fall where they may. Truth is an ally; possible life on what could turn into a slave planet is not.

Sincerely,

Dr. Robert W. Koontz, Ph.D.

Web Site: http://www.DoctorKoontz.com

BIOGRAPHICAL DETAILS OF DR. ROBERT KOONTZ

YAdditional Background Material

I have included below a news article about me when I became employed as a Staff Scientist (II) at the Lawrence Berkeley Laboratory. I have also included a news article about my involvement in an important high energy physics experiment done in the 1975-1978 time frame -- using the electron-positron intersecting storage ring at the Stanford Linear Accelerator Center. That experiment in fact led to discovery of a particle known as the tau lepton. -- for which a Nobel prize was won by Perl in 1995.

With respect to documenting my background, there are also copies of my Ph.D., Master's and Bachelor's degrees shown below, and I have also added a copy of my Honorable Discharge from the US Navy. In addition, there are copies of my last Navy evaluation report from when I worked at the National Security Agency, and there are also evaluation reports from when I worked as an Assistant Professor of Physics at the University of South Dakota.

Other forms of documentation have also been included, such as letters of recommendation from other physicists and a letter of recommendation from Sam Hayes, a friend of the family and then a Pennsylvania State Representative.

I believe that this documentation should prove sufficient to establish my credentials as a physicist beyond a reasonable doubt. That is what I am hoping to do -- as otherwise some of what I have to say concerning such things as antigravity and free energy might be hard for some to believe. Please scroll down.

-- Bob Koontz

The article below is reproduced from the January 14, 1981 Huntingdon Daily News PA newspaper. It describes my research at the Lawrence Berkeley Laboratory, where nine Nobel Prizes have been won. It also mentions my Navy background and other various background credentials.

KOONTZ IS GIVEN DOCTORATE DEGREE
IN NUCLEAR PHYSICS

Robert W. Koontz, Jr., son of Robert and Helen Koontz of 1018 Washington St., Huntingdon, was awarded a doctorate in nuclear physics at the University of Maryland on Dec. 22.

Immediately after graduation he joined the staff of scientists at Lawrence Berkeley Laboratory in California where he is doing experimental research, trying to recreate conditions which are similar to those that were created in nature during the beginning of the universe and trying to create new states of matter that have not been observed before, using conditions of very high pressure and energy.

Koontz was graduation from Huntingdon Area High School in 1965, then enlisted in the U.S. Navy. After boot camp he served aboard the U.S.S. Bayfield until the Navy decided to send him "back to school," where he spent the next two years. During this period he was graduated from 13 schools, some of which include line printer maintenance, college algebra and trigonometry, introduction to computer concepts, elements of computer arts and science, computer maintenance, electronic technical Class A data systems, solid state principals and applications, Fortran programming and petty officer and chief.

His next assignment was with the National Security Agency as an instructor in computer technology at Fort George G. Meade, MD. While on special assignment with NSA, he also attended Johns HopkinsUniversity in Baltimore, studying higher math and physics.

On June 26, 1970, Koontz was given recognition by the National Security Agency for "outstanding academic achievement" when he finished first in a class of 65 at the National Cryptologic School.

In November, 1970, Koontz was named "sailor of the month" and given an award which reads, in part: "This award is given in recognition of his outstanding professional performance, military behavior, leadership and supervisory abilities, military appearance, adaptability and participation in community affairs.

"It is with great pleasure that I present this certificate to one who has set an outstanding example to his shipmates and acted in the highest traditions of the Naval Service." (Signed) Karl B. Kohler, Commanding Officer, Naval Security Group Activity, Fort George G. Meade, Maryland.

After three years as an instructor, Koontz left the Navy with the rank of first class petty officer to continue his education at the University of Maryland. While working for his B.S. degree, he was employed part-time as an instructor in color TV technology and computer programming and maintenance at a private school in Hyattsville, MD. During his last year in undergraduate work at the University of Maryland he was put on the staff as an instructor in physics.

In 1974 Bob was awarded his B.S. degree with honors in physics, then served as a research associate at the university until he joined several teams of scientists at California in order to participate in experiments at Stanford's linear accelerator. Five faculty and staff members from the U. of M., six from Priceton and sic from Pavia, Italy, performed high energy research, investigating the properties of newly-discovered so-called "Strange Particles," Hadron and the "Psi" particles.

In order to study these particles it's necessary to use a high energy linear accelerator. There are only a few in existence that have sufficient power to make these particles. Stanford's accelerator is the largest in the world, extending for two miles, in a perfectly straight line. The particles must be made to collide with other particles at nearly the speed of light to better understand how they interact with other particles. Some of the related experiments consisted of working with electrons and their anti-particles, "Positrons."

After completing this phase of their work Koontz returned to the University of Maryland where he began work on his master's degree in physics, which he received from the same university in 1978. During the time he studied for his master's at the university he worked as an assistant teacher in the field of engineering, which covered hydraulics, mechanics, electrostatics, electrodynamics, magnetism and particle interaction. Much of his studies during this time consisted of work with lasers and the large cyclotron at the university.

Inlieu of a thesis, Koontz has passed a Ph.D. qualifying eamination in 1977, which allowed him to receive his mater's degree on May 19, 1978.

Working toward his Ph.D., Koontz collected over 30,000,000 (30 million) pieces of data, requiring over 1000 hours of computer time (one computer, working steadily, on this problem alone, to the exclusion of all else, for 1000 hours), which lead to a very clear understanding of the results of particle interaction at the high energy levels being utilized. In addition to his other work, Koontz has done research in the field of artificial intelligence (simulation of human thought).

Some of his recent professional articles published in Physical Review Letter and Journal of Physical Review, in collaboration with fellow scientists include: "Alpha Particle Breakup at Incident Energies of 20 and 40 MeV-Nucleon," "Projectile Fragmentation Processes in 35 MeV-amu (a.xy) Reactions," "The Dominant Reaction Mechanisms of 100 MeV Light Ions," "Alpha-Particle Breakup into Multiple Fragments," "Coincidence Studies of Reaction Mechanisms Associated with Non-Equilibrium Particle Spectra," "Multiplicity Measurements of Evaporated Charged Particles in Particle-Particle Coincidence Experiments," "Proton and Neutron Inclusive Spectra and the Importance of the single Nucleon-Nucleon Scattering Process," "Energy Dissipation Process for 100 MeV Protons and the Nucleon-Nucleon Interactions in Nuclei," "'Convrt': A Data Analysis Program for One and Two-Dimensional Analyzer Data," and "DPAP: Data Processing and Plotting Program."

The four pictures below relate to senitive work that I did from 1985 until 2002. The mathematics shown involves numeric multi-dimensional integration. That was an extension of work I did at the University of Maryland, the Lawrence Berkeley Laboratory and then the University of South Dakota, where a student and I did numerical integrations in 20 dimensions. In that time period, from 1978 until about 1985, very few people were doing this kind of mathematics – which was, and is, very useful in computer modeling of complex systems. Most of the other subjects listed below I am not free to say much about. During this time, I was head of my own consulting company. I quit this line of work in 2002 when I decided to pursue research relating to antigravity and free energy.

This is an award I received while working at the National Security Agency. I have redacted the name of the course.

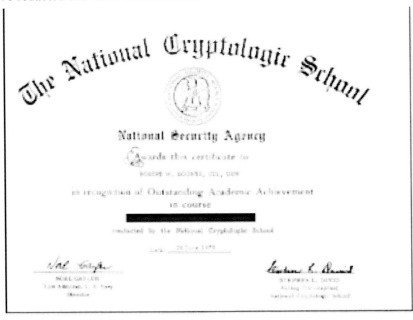

Department of Physics

174 West 18th Avenue
Columbus, OH 43210-1106

Phone 614-292-5713
FAX   614-292-7557

Oct. 17, 2001

To Whom It May Concern:

Dr. Robert W. Koontz was critically instrumental in helping me, a Chinese dissident and citizen of China, who works against the despotic Chinese government, to obtain political asylum in the United States. I will be forever grateful to Dr. Koontz for this, for without political asylum I might have to face 8 years of torture in prison or even death.

Dr. Koontz has also provided me with free web space that I use in trying to fight for democracy in China and to seek the release of those who are wrongly imprisoned by the Chinese government. These include two heroic young men, Yang Zili and Huang Qi. But there are thousands more.

Sincerely,

Yi Ding

Physics Graduate Student, The Ohio State University

APPENDIX ONE

ADDITIONAL SCIENTIFIC EVIDENCE

Report Written by Steven Colbern
CHIEF SCIENTIST FOR
A&S RESEARCH
A 501(C)3 NON-PROFIT ORGANIZATION

A&S RESEARCH
3801 OLD CONEJO ROAD
NEWBURY PARK, CALIFORNIA 91320

Abstract

An object removed from a subjects' toe was found to have produced no significant immune-response or inflammation. Materials analysis revealed the object was composed of carbon nanotubes and crystallites composted of primarily sodium chloride. Recently, carbon nanotubes toxicity has been a concern since its introduction into consumer goods composites. However, it appears objects containing carbon nanotubes may produce a mitigating effect on a subject's immune and inflammatory response. Determination of the underlying cause for such a response by a material could have far reaching implications for organ donation and prosthetic development.

Introduction

The subject of this study noticed a burning pain in the tip of his left, second, toe. Inspection revealed two apparent puncture wounds on the underside of the end of the left, second, toe and a scratch on its right side. One of the puncture wounds in the toe was found to fluoresce green under ultraviolet (UV) illumination.

Over the next four days the pain in the toe increased and felt like a strong electric shock whenever weight was placed on the end of the affected toe. The pain was at a maximum four days after the incident and decreased slowly thereafter.

The subject was examined by a podiatrist, who obtained X-rays of the subject's left foot. A small (~3 mm) foreign object was observed on the X-rays, under the end of the distal phalanx bone of the left, second, toe.

The object resembled a bent piece of wire on the X-ray, but appeared to have approximately the same X-ray density as human bone. A subsequent CAT scan of the left foot confirmed the presence of a foreign object in the same toe.

The object in the subject's toe was surgically removed on September 6, 2008. The object was apparently brittle and broke into 12 pieces during removal. Pathology tests on the tissue surrounding the object showed no inflammation or immunological reaction by the subject's body to the presence of the object. The pieces of the object removed from the subject's toe turned black and then red upon refrigerated storage in blood serum. One of the pieces of the object was used for further materials analysis.

Optical Analysis

The sample used in the optical analysis was a small chunk of solid material, approximately cubic in shape, approximately 1 mm x 0.5 mm x 0.5 mm in dimensions and dark reddish in color. The sample was stored in a small plastic screw-top medical specimen vial and covered in blood serum to prevent degradation until analysis was performed. The sample was visually inspected then imaged under light microscopy using an Olympus dissecting stereomicroscope and an Olympus laboratory microscope. Magnifications from 10x-400x were utilized.[1]

---

1 Magnification is, technically, defined as the ratio of an image metric presented in a form such as printed or projected media to the actual sample metric. However, for the sake of simplification, the magnification in this brief is the objective power (5X, 10X, 20X, 50X, and 100X) multiplied by a standard microscope eyepiece power of 10X.

The sample was first placed under the dissecting microscope to obtain low-magnification images (40x) of the entire piece. The sample was imaged both in the original container, in blood serum, and in the open air (Figures 1a and 1b). Drying the sample appeared to cause no degradation.

figure 1: Sample in air at 40x magnification (a and b), and at 100x (c) and 200x (d). The low-magnification images of the sample revealed that the sample had a somewhat rough and irregular surface. A reddish patina extended over a large percentage of the sample surface, which had a color resembling that of iron oxide.

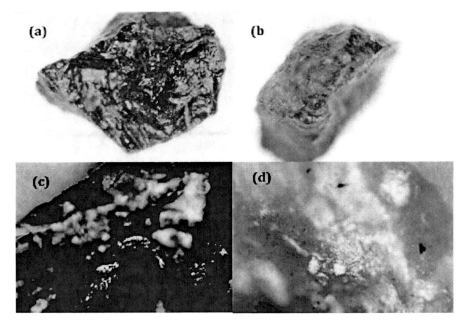

Scanning Electron Microscopy Analysis

SEM imaging was performed using gradually increasing magnification, with the first images taken at a magnification low enough to show the samples bulk structure (Figures 2a and 2b). These images revealed the shiny opalescent phase seen in the light microscope images appeared to represent an outer layer or coating on the sample (Figure 2b). The inner, bulk, portion of the sample consists of a darker[2] material.

2 "Dark", in the context of SEM imaging refers to a surface which absorbs electrons efficiently. Light areas in SEM images are those which reflect electrons efficiently. In this case, the darker material, seen in SEM images, was the Fe-Ni phase as determined by EDX.

Figure 2: Low magnification images of sample at (a)[3] 40x and (b) 230x, showing entire sample and outer layer over darker bulk material.

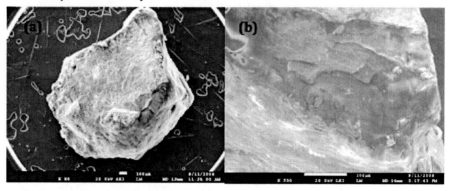

Figure 3: High magnification views of cracks in shiny layer of sample – showing nanofibers (a) 40,000x and (b) 75,000x. Nanofibers with a primary bundle diameter of approximately 10 nanometers (nm) were observed in both the sample outer coating and in the inclusions of light material in the dark areas of the sample (Figures 3a and 3b). These nano-fibers resemble bundles of single-walled carbon nanotubes.

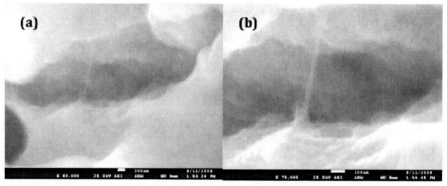

Highly regular crystal inclusions and small pits, both ~500 nm in largest dimension, were also seen in the dark areas of the sample (Figure 4a through 4c). These nanocrystals consisted primarily of sodium and chlorine with traces of sulfur and potassium. The orthorhombic crystal habit is unusual for a sodium chloride crystal which has a cubic crystal structure. Though studies demonstrated the NaCl can form an octahedral habit through solvent modification, it has not been reported that an orthorhombic habit can be obtained.[4] An orthorhombic habit

---

3 Figure 2a shows the shiny sample surface, seen in light microscopy (upper sample surface in image in Figure 1a).

4   *Surf. Sci.* **2003**, *523*, 307-315

from sodium chloride is unlikely, even with solvent modification, since the symmetry and energy of the NaCl primary crystal faces are identical. Typically, NaCl grows in a cubic habit, slightly modified by inhibited growth from the face that is blocked by the surface on which the crystal is growing.

Figure 4: Unusual surface structures in bulk material of sample. (a) nanocrystals are observed in the center of the image, (b) a higher magnification of the crystals reveal orthorhombic crystal habit, and (c) a close-up of the pit seen on the crystal in (b). Notice the pit reveals what appears to be a hollow.

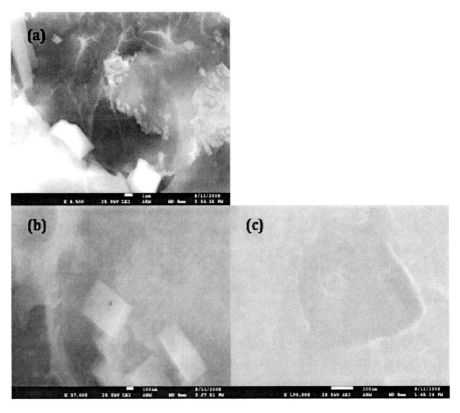

Figure 5 (over): S.E.M. images of object (a) showing microcavities and a long nanoribbon structure, (b) close-up of the nanoribbon structure, (c) network structure and open end of a microcavity structure (top-middle of image), and (d) close-up of a microcavity.

The features in Figure 5 are at the lower diameter range for capillary blood vessels. The large number of these on the surface of the implant indicates the possibility of enhanced biocompatibility. The lack of an inflammatory response while the object was still in the subject supports the conclusion these network may have been capillary blood vessels. A close-up of one of these microcavities in Figure 5d shows the interior walls appear to be very smooth and planar. How these structures might have formed naturally and the purpose of such cavities is uncertain.

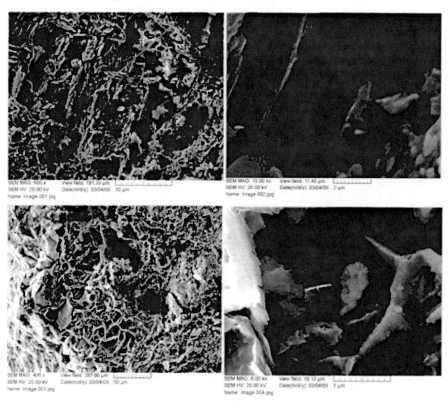

The results of the 532 nm and 633 nm excitation wavelength Raman analyses are shown in Figures 6 and 7. Raman spectra of the sample are compared to Raman spectra of iron oxide, a sample of iron meteorite, and single-walled carbon nanotubes.

Three peaks are observed in the sample's high wavenumber portion of the 532 nm Raman spectrum (1312.0 cm^{-1}, 1566.8 cm^{-1}, and 1587.7 cm^{-1}, Figure 6). These peaks match the single-walled carbon nanotube 532 nm Raman spectrum D-band, metallic (Met), and semi-conducting (SC) G-band peaks. The sample peaks are up-shifted by approximately 3 cm^{-1} to 4 cm^{-1} relative to the single-walled carbon nanotube spectrum.

Figure 6: 532 nm Raman spectrum of sample (top horizontal line) compared to iron oxide from stainless steel (bottom thick line), a naturally occurring iron oxide (middle thick line), and single-walled carbon nanotubes (bottom, thin line that spikes).

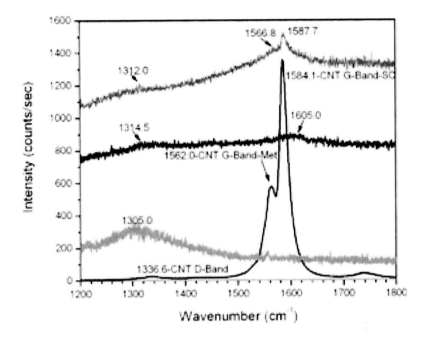

If the sample's peaks are produced by the presence of carbon nanotubes, it is likely the 159 cm-1 peak is a carbon nanotube radial breathing mode (RBM) band. The 633 nm Raman spectrum of the sample (Figure 7) shows the same peak pattern found in the 532 nm spectrum with somewhat greater intensities for RBMs in the lower wavenumber spectral region. The single-walled carbon nanotube G-Band is somewhat less intense.

Figure 7 (over): 633 nm Raman spectrum of sample (top line) compared to iron oxide from stainless steel (middle line), and single-walled carbon nanotubes (bottom thin line).

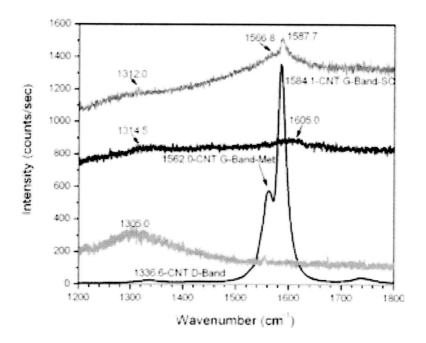

Elemental Analysis

Inductively-coupled plasma-mass spectroscopy (ICP-MS) analysis was performed on a piece of the object removed from Mr. Smith after the SEM, EDX, and Raman data had been obtained. The ICP-MS analysis was performed by an independent laboratory. The sample used for ICP-MS analysis was the same analyzed in the optical, S.E.M., EDX, and Raman analysis.

The sample was digested in a mixture of nitric and hydrochloric acids[5], and an aliquot of the liquid analyzed by ICP-MS. A portion of the sample, which did not dissolve, proved to contain carbon nanotubes by Raman analysis.

The ICP-MS elemental analysis supported EDX results concerning the major components of the sample and also found many trace elements which were not detected by EDX (Table 1). A total of fifty one (51) elements were detected in the sample by ICP-MS.[6]

---

5 See bottom of next page for details of the digestion process.
6 The ICP-MS analysis was not sensitive to halogens (F, Cl, Br, I), or sulfur (S).

The major components of the sample, in order of abundance, were iron (greater than 46%)[7], and nickel (5.20%).

Discussion

The high percentage of iron observed in the chemical analysis data strongly indicates that the red patina on the sample, as delivered, was hydrated iron oxide (rust), a corrosion product formed by oxygen contact with the iron in the sample and the water present in the blood serum in which the sample was stored. The salts dissolved in the blood serum undoubtedly accelerated the corrosion of the metallic portion of the sample.

The black material, seen on the sample pieces by Dr. Leir, soon after removal from the patient, was freshly formed iron oxide[8], in which hydration had not yet been completed. The sample had an outer coating of a non-metallic, ceramic-like material, which was approximately 100 nm-200 nm in thickness. This material had a somewhat rough texture, as seen under SEM, with surface irregularities up to several microns in size. Large numbers of inclusions of what appeared to be the same material[9], which were typically several microns in size, were also seen in the metallic phase.

The high concentration of non-metallic inclusions in the metallic phase probably account for the brittleness of the original object. Inclusions of unlike material in this size range, which do not bind well with the sample matrix, act as points of stress concentration during episodes of mechanical action, leading to cracking at much lower stress levels than would be the case with a homogeneous metallic material. The presence of these inclusions is the most likely cause of the breakage of the original object into small pieces, during its surgical removal.

The non-metallic, ceramic-like material contains mainly carbon (C), oxygen (O), silicon (Si), sulfur (S), aluminum (Al), calcium (Ca), iron (Fe) and nickel (Ni), with smaller amounts of sodium (Na), phosphorus (P), chlorine (Cl), potassium (K), and titanium (Ti), and chemically

---

7 An exact percentage of iron cannot be obtained from this analysis, because the mass spectrometer detector was saturated. The EDX-derived value for the percentage of iron in the sample (~94%) is more accurate, in this case.

8 This material (Fe_2O_3) is black in color, and turns red after long exposure to water.

9 EDX elemental analyses of the inclusions and the outer coating were very similar.

resembles a biological hard part, such as shell or bone[10]. The similarity in composition of the non-metallic phase to biological material may be responsible for the lack of immune response to the object by the patient's body.

The opalescence of this material, seen in the light microscopy images, indicates the presence of an organized, layered structure, such as occurs in mother-of-pearl or opal, which reflects and refracts light strongly into different color bands.

The Raman data, showing what appears to be carbon nanotube D- and G- bands, along with possible radial breathing modes, strongly indicates the presence of carbon nanotubes (CNTs). This is confirmed by the SEM images, which show bundles of nanotubes, with high carbon content (EDX data), which appear nearly identical to SEM images of commercial arc-process, single-walled, CNTs.

Therefore, the data indicates the majority of the non-metallic phase material is probably composed mainly of carbon nanotubes, which are covered and/or filled by a shell-like coating of aluminum, calcium, iron, nickel, and titanium silicates, oxides, sulfates, and phosphates.

A smaller percentage of the non-metallic phase of the sample is composed of the very regularly sized (500 nm), and shaped, sodium, potassium, and iron, chloride and sulfate-containing crystals, seen in the SEM images.

These crystals appear to be quite regular in size and shape and possess an orthorhombic habit. Though the blood serum in which the sample was stored certainly contained sodium chloride, the orthorhombic habit is unusual for a sodium chloride crystal which has a cubic close packed unit cell and should primarily have a habit that is cubic. Though habit modification in crystals through crystal face poisoning with compounds different from the primary growth material is well documented in peer reviewed literature, it is not possible in the case of a simple crystal structure such as sodium chloride, which has equivalent surface energies on the dominant growth faces.

---

10   Elemental composition of the non-metallic, ceramic-like phase of the sample was derived from EDX data.

Conclusions

The sample consists mainly of iron, with a high carbon and oxygen content. The iron base material contains 5.2% nickel, and is highly magnetic. The sample consists of two major phases: an iron/ nickel (Fe/Ni) phase and a non-metallic phase resembling a hard biological substance such as shell, tooth, or bone. The iridescence of the non-metallic phase, seen in light microscopy, suggests a layered microstructure similar to mother-of-pearl or opal.

The similarity of the non-metallic phase composition to biological material may be responsible for the patient's lack of immune response to the object. This non-metallic phase is high in carbon, oxygen, silicon, magnesium, aluminum, sulfur and phosphorus, and is present as an outer covering on the sample and as inclusions in the metallic Fe/Ni phase. The non-metallic phase of the sample also contains bundles of carbon nanotubes, perhaps covered or filled, with calcium and magnesium silicates, phosphates and sulfates.

The lack of immune-response from this sample begs further and more detailed analysis of samples with similar composition. Specifically, kinetic studies of such materials in blood serum would assist in determining the mechanism by which the material minimized rejection by the body.

Table 1: **ICP-MS** Results of sample piece.

Trace impurities by SCOP7040, Rev 9
inductively coupled plasma – mass- spectrometry

Sample ID; SN Chunk 080911

	ppm	Detection Limit		ppm	Detection Limit
Aluminum	260	30	Molybdenum	9.3	0.05
Antimony	0.37	0.2	Neodymium	0.39	0.02
Arsenic	17	0.4	Nickel	52000	0.1
Barium	96	0.1	Niobium	0.37	0.1
Beryllium	ND	0.05	Osmium	2.2	0.09
Bismuth	ND	0.03	Palladium	3.3	0.02
Boron	15	3	Phosphorous	1600	10
Bromine	ND	5	Platinum	10	0.02
Cadmium	ND	0.09	Potassium	ND	50
Calcium	1500	30	Praseodymium	0.11	0.02
Cerium	0.85	0.03	Rhenium	0.66	0.02
Cesium	ND	0.02	Rhodium	2.8	0.02
Chromium	13	0.2	Rubidium	0.15	0.02
Cobalt	2200	0.09	Ruthenium	8.0	0.02
Copper	170	0.3	Samarium	0.13	0.02
Dysprosium	0.11	0.02	Selenium	2.5	0.02
Erbium	0.07	0.02	Silicon	2700	50
Europium	0.03	0.02	Silver	230	10
Gadolinium	0.13	0.02	Sodium	230	10
Gallium	130	0.02	Strontium	0	0.2
Germanium	300	0.1	Tantalum	ND	0.07
Gold	0.90	0.09	Tellurium	ND	0.1
Hafnium	0.10	0.02	Thallium	ND	0.2
Holmium	ND	0.02	Thorium	0.23	0.02
Iodine	ND	0.9	Thulium	ND	0.02
Indium	3.6	0.05	Tin	6.5	0.1
Iron	460000	4	Titanium	20	0.3
Lanthanum	ND	1	Tungsten	1.9	0.07
Lead	1.3	0.1	Uranium	0.21	0.02
Lithium	ND	1	Vanadium	21	1
Lutetium	ND	0.5	Ytterbium	0.05	0.02
Magnesium	890	5	Yttrium	0.05	0.4
Manganese	62	0.1	Zinc	44	2
Mercury	ND		Zirconium	4.4	0.3

Note: the entire sample (0.0058 g) was mixed with 0.5 mL nitric acid and 0.5 mL hydrochloric acid and heated on a hot block set at 110 degrees C for 1 hour. The mixture was cooled, 0.5 mL 30% hydrogen peroxide added, and heated 30 min. A portion remained undissolved, therefore, silicon or other elements that have low solubility in this acid mixture may be biased now. The solution was mixed with internal standards, diluted to 10 g and a 1 : 100 dilution also analyzed by ICPMS.
Date analyzed: 11-26-08
Elements not analyzed: all glasses, C, S, Sc, In, Tb

ANALYSIS PERFORMED BY STEVE COLBERN B.S. M.S. CHIEF SCIENTIST, A&S RESEARCH

APPENDIX TWO
SCIENTIFIC DATA

The following pages contain academic scientific data for your perusal and information. Some of this may not be useful for the average reader, but those familiar with more advanced scientific data will understand its importance. Please note that what appears in this one chapter only covers a very superficial amount. If you hold a scientific degree and would like to discuss particular details of the data, please feel free to contact us through our webmaster, Gary Lowery at alienscalpel.com. In turn, we will put you in touch with Steve Colbern, our chief scientist with A&S Research.

EDX Photos under High Magnification

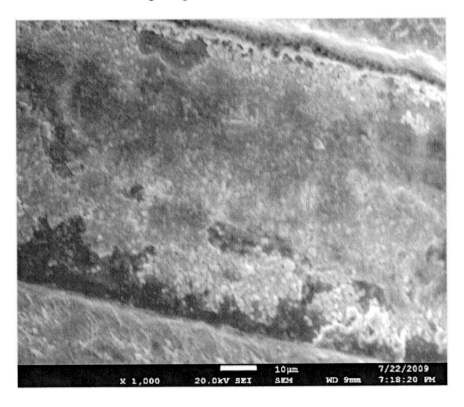

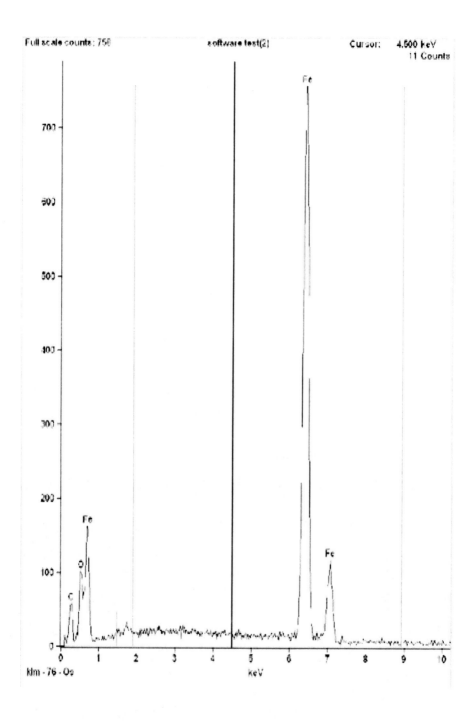

software test(3)

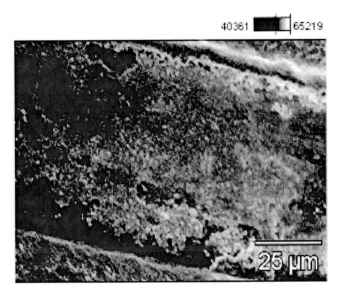

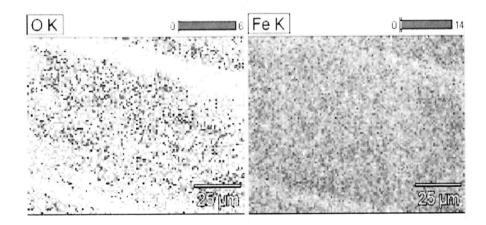

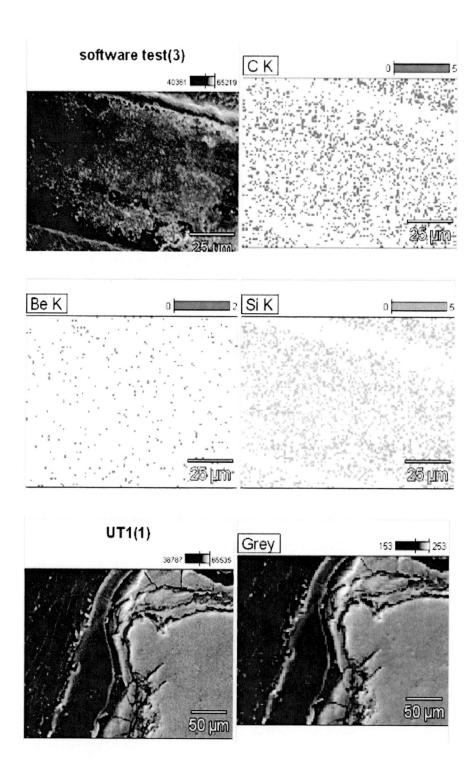

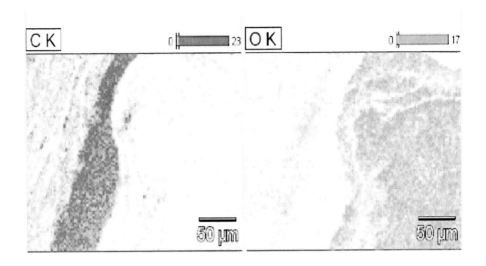

Data Type Counts   Mag 400   Acc. Voltage 20.0 kV

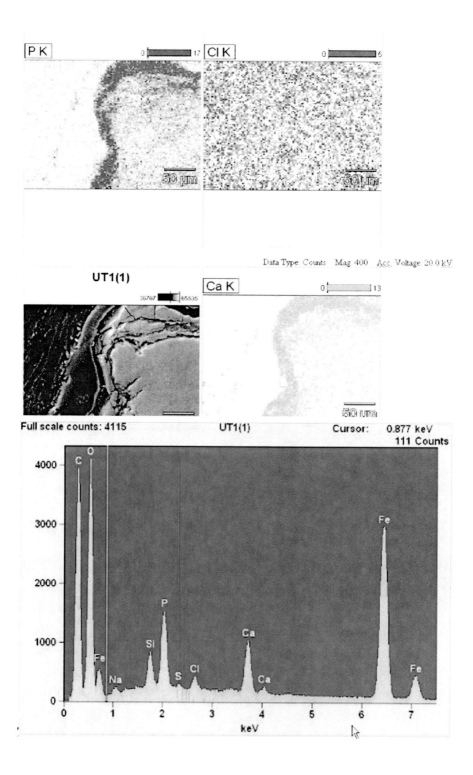

UT1(2)

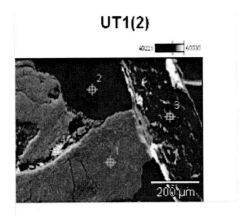

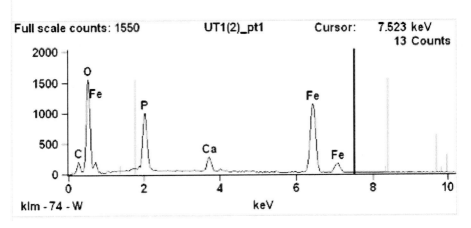

Full scale counts: 1550   UT1(2)_pt1   Cursor: 7.523 keV
13 Counts

klm - 74 - W

keV

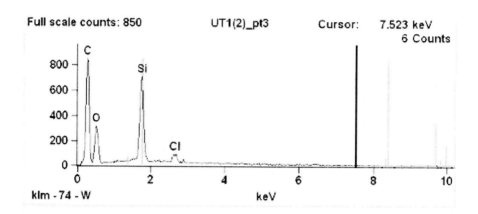

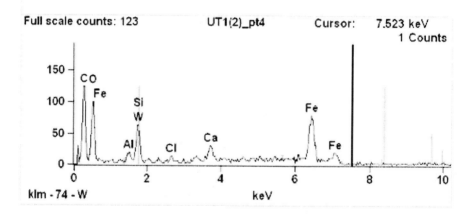

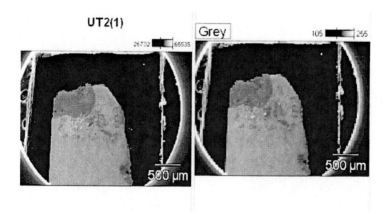

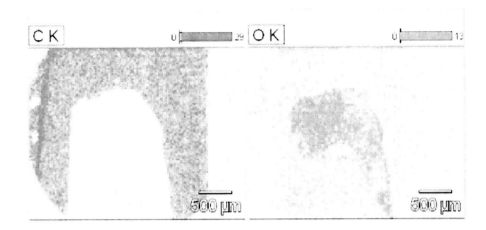

Data Type Counts   Mag 35   Acc. Voltage 20.0 kV

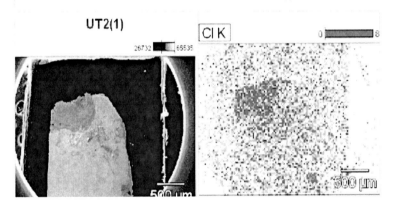

UT2(1)

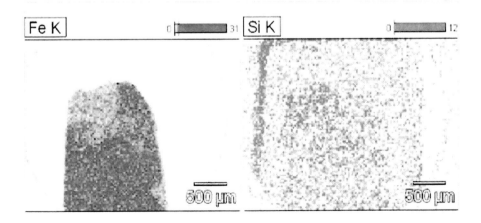

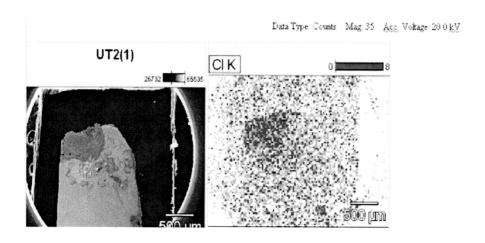

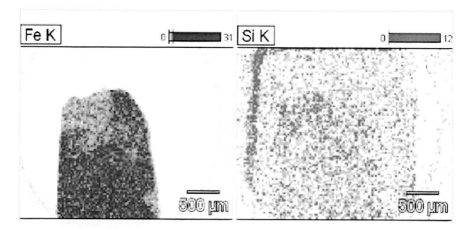

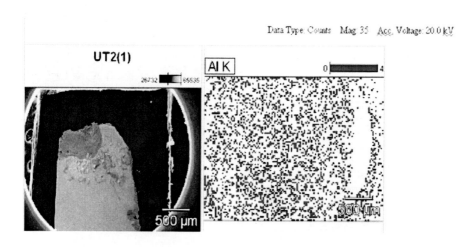

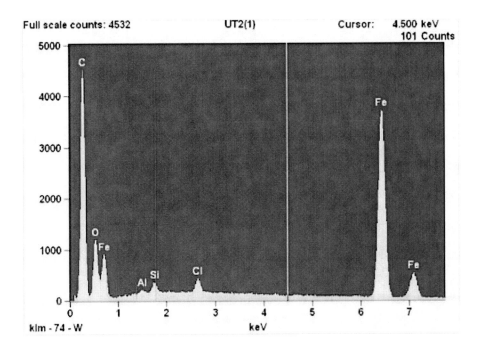

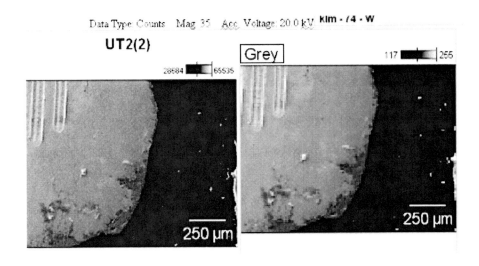

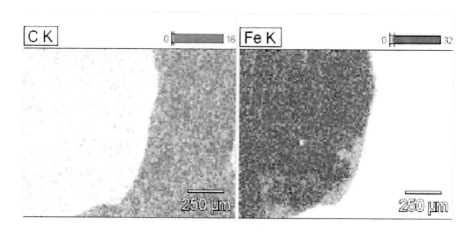

Data Type: Counts   Mag: 80   Acc. Voltage: 20.0 kV

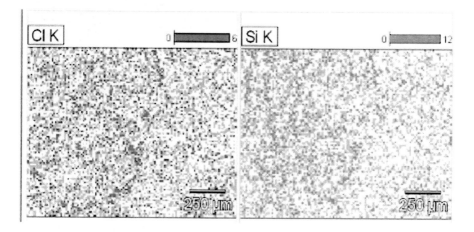

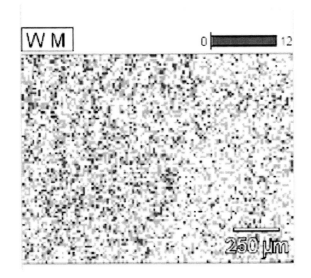

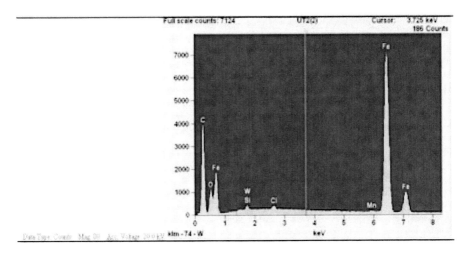

UT2(3)

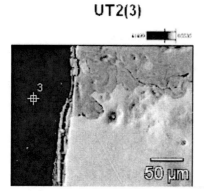

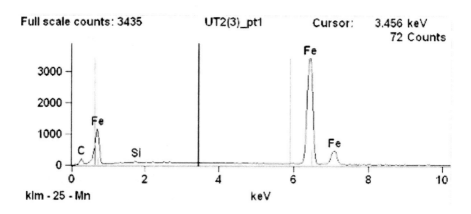

Full scale counts: 3435   UT2(3)_pt1   Cursor: 3.456 keV
                                                               72 Counts

klm - 25 - Mn

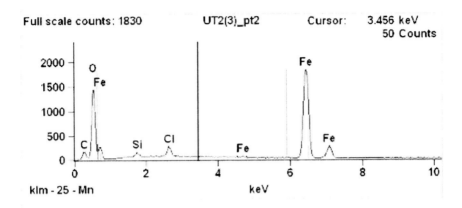

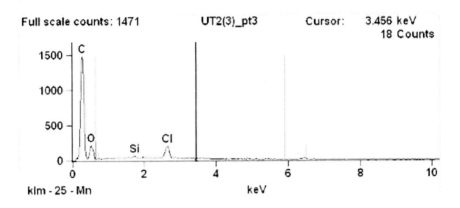

SPECIMAN (JOHN SMITH – STEVE)

8 February, 2011

Summary of Characteristics of UT Implant Sample

This sample was removed from the right thigh of one of the patients researched by A&S Research and analyzed at the University of Toronto.

A second analysis of the object was done in 2009, by Mr. Steve Colbern. In the second analysis, the object was imaged by light microscopy, scanning electron microscopy (SEM), and subjected to energy dispersive X-ray (EDX) elemental analysis. Elemental mapping EDX analysis was used, in addition to the standard area method, to determine the uniformity of the distribution of elements in the sample.

Raman spectroscopy was also done on the sample to detect the possible presence of organic material, and/or carbon nanotubes. Carbon nanotubes were previously detected in at least two other implant samples.

The object appeared to be manufactured, and showed several sets of parallel grooves, as well as an outer layer of material. This outer layer (see image) contained much more carbon than the interior.

Iron was the largest component element in the sample, and the object appeared to be composed of an iron alloy. In addition the type of iron seen in this sample was not similar to the other implant iron as there was no Nichol content to indicate the presence of Meteoric iron.

Other elements detected in the object by EDX, roughly in order of abundance, included carbon (C), oxygen (O), phosphorus (P), calcium (Ca), silicon (Si), sulfur (S), chlorine (Cl), sodium (Na), aluminum (Al), manganese (Mn), beryllium (Be), and tungsten (W). A total of 13 elements were detected by EDX.

EDX mapping showed that the metallic elements in the interior were distributed very homogeneously. This was not true, however, of the non-metallic elements in the object, and there appeared to be a coating on certain areas of the metal, composed mainly of calcium phosphate, calcium sulphate, and calcium silicate.

The Raman data showed definitive evidence of the presence of carbon nanotubes. Organic material was seen in another related sample. It should also be noted that the analysis performed by the University of Toronto was not specifically looking for Carbon Nano structures. It should also be noted that a number of other unknown structures were seen during the EDX and SEM examinations such as Ovoids, Spheres and Crystals.

The metallic portion of this sample was also covered with a biological coating which was analyzed separately.

Raman Spectroscopy Report on Implant Objects TC-1 and TC-2

SPECIMAN (John Smith-Steve) 8 February, 2011

Experimental Procedure

The TC-1 and TC-2 implant objects were subjected to Raman spectroscopy, using a Horiba micro Raman spectrometer, using the 10X objective. The 532 nm (green) diode laser was used as the excitation source.

Spectra from several regions of each object were taken to investigate the degree of homogeneity in the structures of the objects. The spectra were obtained between 100 cm^{-1} and 1800 cm^{-1}, so as to view the region of the Raman spectrum which is most likely to reveal the presence of Raman scattering bands related to the presence of carbon nanotubes.

Results

The Raman spectra of both the TC-1 and TC-2 objects show a strong luminescence background, as is typical of the Raman spectra of metallic objects.

The spectra of both objects also show a potential single-walled carbon nanotube (SWCNT) G-Band Raman scattering [1] near 1600 wavenumbers (cm^{-1}).

A potential carbon nanotube D-Band, near 1350 cm^{-1}, is also seen in the spectra of both objects, which is more intense in the spectrum of object TC-2 than in TC-1 (Figures 1 and 2 and Figures 6 and 7).

A 532 nm Raman spectra of commercial SWCNT, and the "John Smith" implant object, which was determined to contain SWCNT, are shown, for comparison, in Figure 3. Both of these samples display Raman G and D bands very close to the wavenumber values of the possible G and D band peaks seen in the TC-1 and TC-2 objects.

The spectra of the TC objects also show four Raman bands which are consistent with those of the radial breathing mode [2] (RBM) bands of SWCNT (Figures 4 and 5). These bands average approximately 155 cm^{-1}, 220 cm^{-1}, 295 cm^{-1}, and 410 cm^{-1}, which would correspond to SWCNT diameters of approximately 1.6 nanometer (nm) [3], 1.1 nm, 0.8 nm, and 0.6 nm (see Table 1).

The SWCNT RBM bands in the John Smith sample were at 217 cm^{-1}, 289 cm^{-1}, and 394 cm^{-1}, corresponding to SWCNT diameters of 1.14 nm, 0.86 nm, and 0.63 nm. These bands were especially intense in the sample after treatment with acid to remove the metal matrix from the SWCNT. The commercial SWCNT sample had an average diameter of approximately 1.6 nm.

Drawing of Magnetic Anomalies

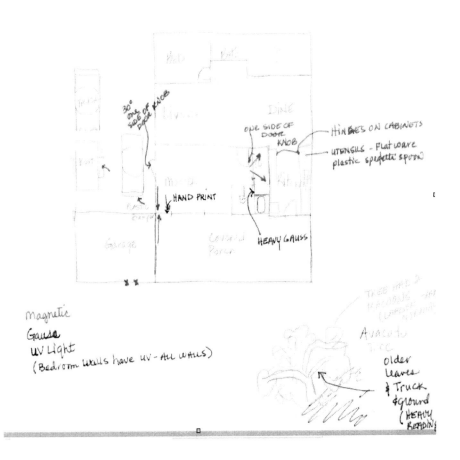

Steve's typical Southern California home with the usual tarpaper, single story construction.

Soil under Steve's avocado tree would undergo spontaneous combustion. This would happen directly under the tree area but the surrounding soil also showed extreme high readings. Comparative studies showed abnormal amounts of Bromine. The tree had branches that indicated an excess of 10 Mil Gauss registering on the Gauss meter.

Soil under Avocado tree.

Steve had a boat with a fiberglass hull. This was also magnetized.

WORD MEANINGS AND SEMANTIC CONCLUSIONS

The following are official meanings of the words and associated words found:

UNIDENTIFIED FLYING OBJECT

Collins World English Dictionary:

unidentified (ʌnaɪ,dɛnt'faɪd) *adj* **unidentifiable** not identified or recognized: *an unidentified man*

unidentified If you describe someone or something as unidentified, you mean that no-one knows who or what they are.

Merriam-Webster Dictionary

Unidentified: 1

known but not named <an unidentified worker reported the security breach>

Synonyms

anonymous, given, one, some, unidentified, unnamed, unspecified

Related Words

particular, specific

Near Antonyms

known, named, specified

Unidentified: 2

not named or identified by a name <some unidentified person helped them and then left quietly>

Synonyms

anonymous, faceless, incognito, innominate, unbaptized, unchristened, unidentified, unnamed, untitled

According to the **American Heritage Dictionary**, the word **Identified means the following:**

To ascertain or establish the identity of,

To be or cause to be identical, regard as the same,

To associate; connect

FLYING

adjective

1. making <u>flight</u> or passing through the air; that <u>flies</u>: *a flying insect; an unidentified flying <u>object</u>.*

2. floating, fluttering, waving, hanging, or moving freely in the air: *flying banners; flying hair.*

Relevant Questions

What is a Fly?

What is the Scientific Name?

What Makes Rockets Fly?

What is Needed to Fly?

noun

The act of moving through the air on wings; flight.

Flying is always a great word to know.

So is flibbertigibbet.

So is lollapalooza.

So is quincunx. Do these all mean?

(a chattering or flighty, light-headed person)

(an extraordinary or unusual thing, person, or event; an exceptional example or instance)

(a fool or simpleton; ninny)

(an extraordinary or unusual thing, person, or event; an exceptional example or instance)

(the offspring of a zebra and a donkey)

(an arrangement of five objects, as trees, in a square or rectangle, one at each corner and one in the middle)

adverb

Nautical, without being fastened to a yard, stay, or the like: a sail set flying.

Origin: before 1000; Middle English (**noun**); Old English flēogende (**adj.**). See fly, -ing

Related forms

non·fly·ing, adjective.

un·fly·ing, adjective.

Related Words for: Flying

flight, fast-flying, aflare, flaring, waving

Dictionary.com Unabridged

FLY

verb (used without object)

1. to move through the air using wings.

2. to be carried through the air by the wind or any other force or agency: bits of paper   flying about.

3. to float or flutter in the air: flags flying in the breeze.

4.to travel in an aircraft or spacecraft.

5. to move suddenly and quickly; start unexpectedly: He flew from the room.

6. to change rapidly and unexpectedly from one state or position to another: The door   flew open.

7. to flee; escape.

8. to travel in space: The probe will fly past the planet.

9. to move or pass swiftly: How time flies!

10. to move with an aggressive surge: A mother fox will fly at anyone approaching her kits.

11. Baseball.

a. to bat a fly ball: He flied into right field.

b. to fly out.

12. Informal: to be acceptable, believable, or feasible: It seemed like a good idea, but it just wouldn't fly.

verb (used with object)

13. to make (something) float or move through the air: to fly a kite.

14. to operate (an aircraft, spacecraft, or the like).

15. to hoist aloft, as for display, signaling, etc.: to fly a flag.

16. to operate an aircraft or spacecraft over: to fly the Pacific.

17. to transport or convey by air: We fly merchandise to Boston.

18. to escape from; flee: to fly someone's wrath.

19. Theater.

a. to hang (scenery) above a stage by means of rigging supported by the gridiron.

b. to raise (scenery) from the stage or acting area into the flies.

noun

20. a strip of material sewn along one edge of a garment opening for concealing buttons, zippers, or other fasteners.

21. a flap forming the door of a tent.

22. Also called tent fly: a piece of canvas extending over the ridgepole of a tent and forming an outer roof.

23. an act of flying; a flight.

24. the course of a flying object, as a ball.

25. Baseball: fly ball.

26. British: a light, covered, public carriage drawn by one horse; hansom; hackney coach.

27. Machinery: a horizontal arm, weighted at each end, that pivots about the screw of a press so that when the screw is lowered the momentum of the fly will increase the force of the press.

28. Also called fan. Horology: a regulating device for chime and striking mechanisms, consisting of an arrangement of vanes on a revolving axis.

29. Printing.

a. (in some presses) the apparatus for removing the printed sheets to the delivery table.

b. Also called flyboy. (formerly) a printer's devil employed to remove printed sheets from a press.

30. the horizontal dimension of a flag as flown from a vertical staff.

b. the end of the flag farther from the staff. Compare hoist (def 7).

31. flies, Also called fly loft. Theater: the space above the stage used chiefly for storing scenery and equipment.

32.Nautical: a propeller like device streamed to rotate and transfer information on speed to a mechanical log.

Verb phrases

33. fly out, baseball, softball. To be put out by hitting a fly ball that is caught by a player of the opposing team.

Idioms

34. fly blind. blind (def 33) .

35. fly in the face of, to act in defiance of (authority, custom, etc.). Also, fly in the teeth

36. fly off the handle. handle (def 16).

37. go fly a kite, Slang.

a. to put up with or get used to matters as they stand.

b. to confine oneself to one's own affairs.

c. to cease being a nuisance: If she gets mad enough she'll tell me to go fly a kite.

38. let fly,

a. to hurl or propel (a weapon, missile, etc.).

b. to give free rein to an emotion: She let fly with a barrage of angry words.

39.on the fly,

a. during flight; before falling to the ground: to catch a baseball on the fly.

b. hurriedly; without pausing: We had dinner on the fly.

Origin: before 900; Middle English flīen, Old English flēogan; cognate with Old High German

fliogan, German fliegen, Old Norse fljuga

Related forms

fly·a·ble, adjective.

fly·a·bil·i·ty, noun.

non·fly·a·ble, adjective.

re·fly·a·ble, adjective.

un·fly·a·ble, adjective.

Synonyms
1. Fly, flit, flutter, hover, soar refer to moving through the air as on wings. Fly is the   general term: Birds fly. Airplanes fly. To flit is to make short rapid flights from place to place: A bird flits from tree to tree. To flutter is to agitate the wings tremulously, either without flying or in flying only short distances: A young bird flutters out of a nest and in again. To hover is to linger in the air, or to move over or about something within a narrow area or space: hovering clouds; a hummingbird hovering over a blossom. To soar is to (start to) fly upward to a great height usually with little advance in any other direction, or else to (continue to) fly at a lofty height without visible movement of the wings: Above our heads an eagle was soaring.

Dictionary.com, Unabridged. Based on the Random House Dictionary, © Random House, Inc. 2012.

Collins World English Dictionary

Adj

(prenominal) hurried; fleeting: a flying visit

(prenominal) designed for fast action

(prenominal) moving or passing quickly on or as if on wings: a flying leap ; the flying hours

hanging, waving, or floating freely: flying hair

nautical (of a sail) not hauled in tight against the wind

the act of piloting, navigating, or traveling in an aircraft

(modifier) relating to, capable of, accustomed to, or adapted for flight: a flying machine

Related: Volant

Etymonline

Word Origin & History

Flying

O.E. fleogende, prp. of fly (v.1). Flying buttress is from 1660s; flying fish is from 1510s.

Flying saucer first attested 1947, though the image of saucers for unidentified flying

objects is from at least 1880s. Flying Dutchman ghost ship first recorded c.1830, in

Jeffrey, Baron de Reigersfeld's "The Life of a Sea Officer." Flying colors (1706)

probably is from the image of a naval vessel with the national flag bravely displayed.

Online Etymology Dictionary, © 2010 Douglas Harper

FLYING

in animals, locomotion of either of two basic types-powered, or true, flight and gliding. Winged (true) flight is found only in insects (most orders), most birds, and bats. The evolutionary modifications necessary for true flight in warm-blooded animals include those of the forelimbs into wings; lightening and fusion of bones; shortening of the torso; enlargement of the heart and thoracic muscles; and improved vision. Similar modifications in insects have occurred through different evolutionary pathways. The advantages conferred by flight are also great: in terms of numbers of species as well as numbers of individuals, insects, birds, and bats are among the most successful animal groups.

Encyclopedia Britannica, 2008. Encyclopedia Britannica Online.

Example sentences

Frivolous lawsuits, intimidation, mobbing are not the flying buttresses of modern science.

Moths, on the other hand, require a stream of electrical signals in order to keep flying.

It is about a bird who raises his consciousness until he can perform such paranormal feats as flying through solid rock.

As publishing struggles, the memoirs of stand-up comedians are flying off the shelves.

Freshly made food, piled in colorful mountains on giant platters, is flying by on its way upstairs.

There were still bees flying, even in the rain, but not many.

In fact, aircraft often trigger lightning when flying through a heavily charged region of a cloud.

Flying is inherently not green and never will be green.

But of course it has also to some degree demonstrated that pilots are increasingly unimportant in the flying of the machine.

She tore down the runway toward him with open arms, her spirits- and feet-flying.

Related Words

Flight, flying, jenny, kite flying, wing, air-to-surface, color

constructional homonymity

Distinguished Flying Cross, dodger, flying circus, flying colors, flying dragon

Matching Quote

1. extending through the air.

2 moving swiftly.

3. made while moving swiftly: *a flying leap.*

4. very hasty or brief; fleeting or transitory: *a flying visit; a flying remark.*

designed or organized for swift <u>movement</u> or action.

<u>fleeing</u>, running away, or taking <u>flight</u>: *They pursued the flying enemy.*

Nautical: (of a sail) having none of its edges fastened to spars or stays.

Next, let's look at the term <u>OBJECT</u>

ob·ject

noun

1. anything that is visible or tangible and is relatively stable in form.

2. thing, person, or matter to <u>which</u> thought or action is directed: *an object of medical investigation.*

3. the end toward which effort or action is directed; <u>goal</u>; purpose: *Profit is the object of business.*

4. a person or thing with reference to the impression made on the mind or the feeling or emotion elicited in an observer: *an object of curiosity and pity.*

5. anything that may be apprehended intellectually: *objects of thought.*

Optics: the thing of which a lens or mirror forms an <u>image</u>.

Grammar: (in many languages, as English) a <u>noun</u>, noun phrase, or noun substitute representing by its syntactical position either the goal of the action of a <u>verb</u> or the goal of a preposition in a prepositional phrase, as *ball* in *John hit the ball, Venice* in *He came to Venice, coin* and *her* in *He gave her a coin.* Compare <u>direct object</u>, <u>indirect object</u>.

Computers: any item that can be individually selected or manipulated, as a picture, data file, or piece of text.

Metaphysics: something toward which a cognitive act is directed.

Relevant Questions

What is an Object?

Where Can I Play Hidden...

What is Objective?

Are there Hidden Objects...

verb (used without object)

to offer a <u>reason</u> or argument in opposition.

to express or feel disapproval, dislike, or distaste; be averse.

to refuse or attempt to refuse to permit some action, speech, etc.

Object is one of our favorite verbs.

So is yaff

So is lollygag.

So is subtilize. Do these mean:

> **To: bark, yelp**
>
> **to flee: abscond:**
>
> **to spend time idly; loaf.**
>
> **chat, to converse**
>
> **to introduce subtleties into or argue subtly about**
>
> **chat, to converse**

verb *(used with object)*

to state, claim, or cite in opposition; put forward in <u>objection</u>: *Some persons objected that the proposed import duty would harm world trade.*

Archaic. to bring forward or adduce in opposition.

Origin: (noun) Middle English: something perceived, purpose, objection < Medieval Latin *objectum* something thrown down or presented (to the mind), noun use of neuter of Latin *objectus* (past participle of *objicere*), equivalent to *ob- <u>ob-</u> + jec-* (combining form of *jacere* to throw; + *-tus* past participle suffix; (v.) Middle English *objecten* to argue against (< Middle French *obje* (*c*) *ter*) < Latin *objectāre* to throw or put before, oppose.

Related forms

ob·jec·tor, *noun.*

o·ver·ob·ject, *verb.*

pre·ob·ject, *verb (used without object)*

re·ob·ject, *verb (used with object)*

su·per·ob·ject, *verb (used without object)*

un·ob·ject·ed, *adjective.*

Related Words for: object

Can be confused: abject, object.

Synonyms

1. objective, target, destination, intent, intention, motive. See aim.

physical object, aim, objective, target

View more related words »

Dictionary.com Unabridged

object

1. objection

2. objective

1. a tangible and visible thing

2. a person or thing seen as a focus or target for feelings, thought, etc: *an object of affection*

3. an aim, purpose, or objective

4. *informal* a ridiculous or pitiable person, spectacle, etc.

5. *philosophy* that towards which cognition is directed, as contrasted with the thinking subject; anything regarded as external to the mind, esp in the external world.

6. *grammar* direct object See also indirect object a noun, pronoun, or noun phrase whose referent is the recipient of the action of a verb.

7. *grammar* a noun, pronoun, or noun phrase that is governed by a preposition

8. **no object** not a hindrance or obstacle: *money is no object*

9. *computing* a self-contained identifiable component of a software system or design: *object-oriented programming* from Late Latin *objectus* something thrown before (the mind), from Latin *obicere;*

object 2

verb: (often followed by to)

[tr; takes a clause as object) to state as an objection: he objected that his motives had been good to raise or state an objection (to); present an argument (against) from Latin obicere, from ob- against + jacere to throw.]

Collins English Dictionary - Complete & Unabridged 10th Edition 2009 © William Collins Sons & Co. Ltd. 1979, 1986 © HarperCollins Publishers 1998, 2000, 2003, 2005, 2006, 2007, 2009

Etymonline

Word Origin & History

OBJECT

"tangible thing, something perceived or presented to the senses," from M.L. objectum "thing put before" (the mind or sight), neut. of L. objectus, pp. of obicere "to present, oppose, cast in the way of," from ob "against" + jacere "to throw" (see jet. Sense of "thing aimed at" is late 14c. No object "not a thing regarded as important" is from 1782. Object lesson "instruction conveyed by examination of a material object" is from 1831. "to bring forward in opposition," from L. objectus, pp. of objectare "to cite as grounds for disapproval," freq. of obicere, or else lit. "to put or throw before or against"

American Heritage Cultural Dictionary

object definition

A part of a sentence; a noun, pronoun, or group of words that receives or is affected by the action of a verb. (See direct object, indirect object, and objective case.)

The American Heritage® New Dictionary of Cultural Literacy, Third Edition, Copyright © 2005 by Houghton Mifflin Company. Published by Houghton Mifflin Company. All rights reserved.

OBJECT-ORIENTED

In object-oriented programming, an instance of the data structure and behavior defined by the object's class. Each object has its own values for the instance variables of its class and can respond to the methods defined by its class.

For example, an object of the "Point" class might have instance variables "x" and "y" and might respond to the "plot" method by drawing a dot on the screen at those coordinates. (2004-01-26)

Idioms & Phrases

OBJECT

see *money is no object.*

The American Heritage® Dictionary of Idioms by Christine Ammer. Copyright © 1997. Published by Houghton Mifflin.

Example sentences

In fact, matter as a visible object is of no great use any longer, except as the mold on which form is shaped.

With ram accelerators, diameter's the thing: the bigger the bore, the bigger the object that can be fired out of it.

Thereafter the touch of every object would bring a glowing memory of how that object looked.

One of her functions in this world was to serve as an object lesson for her nephew.

SEMANTIC AND SCIENTIFIC CONCLUSIONS

As the reader can plainly see, these terms do not correctly apply to the UFO phenomena. In the first place, so called UFO's are identified as some type of visual phenomena plainly seen by millions of people and should belong to an unknown category of worldwide, observed visual aberration. They have, without a doubt, been identified as a *visual anomaly*. If we visualize a commercial aircraft, we may not be able to recognize it as a Boeing 747 or other model, but we do recognize it as being in the category of an aircraft and therefore we can say they have been IDENTIFIED. Likewise, UFO's are identified visual phenomena to the point that we know them to be unquestionable anomalies based on current technology.

Next, let us consider the dictionary word, FLYING. There is not one single respected scientist that has studied the UFO subject who can prove that these entities are flying. When they are observed at one point in the sky, it may take only a blink of an eye before they are seen elsewhere at a long distance away from the original sighting area. It is then a foolish scientific assumption to assume the entity flew to the other location when we know that science can demonstrate the almost immediate transference of matter through a time continuum, black hole or inter-dimensional travel. The word "flying" then does not apply to the UFO phenomena.

Last but not least is the term OBJECT. Again, using the strict terminology of dictionary sources we can easily see that the observation of this unknown should not be referred to as an OBJECT. Even to extend our reasoning to believe they are some form of matter that we do not understand, infers a leap of scientific faith that cannot be proven by our chosen scientific methodology.

The conclusion reached by this author is that 1) A worldwide lie has been perpetrated long ago stating that "UFOs do not exist" and is still trying to be maintained. 2) Our language limits us in our assessment of UFOs and, in effect, helps to perpetrate this lie. 3) It is safe to conclude that the preponderance of scientific evidence far outweighs the illogical semantics in order to prove, beyond the shadow of a doubt, that such objects do in fact exist along with their corresponding intelligent life. This intelligent life has left clear, unquestionable proof that they are here – because I have found it.

OTHER BOOKS BY DR. ROGER LEIR

The Aliens and the Scalpel

The personal story of a professional physician's work involving one of the greatest breakthroughs of all time – scientific proof that anomalous bio-electromagnetic implants of extraterrestrial origin have been removed from persons reporting alien abduction experiences. This revised and updated book includes and is supported by new scientific reports and a new photo gallery section. The information presented here provides enough evidence for us to confidently believe that we have cosmic companions and they here with us now. **248 pages • $22.95**

UFO Crash in Brazil

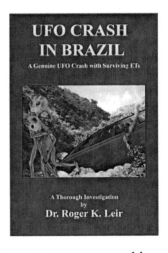

If you think that UFOs are not real or that only crazy people believe in aliens, think again. This story was originally reported in the Wall Street Journal, then was forgotten. According to numerous eyewitnesses, an unidentified flying object crashed near Varginha, Brazil on January 20, 1996 and at least two beings were reported to have survived. One was taken to a local hospital for treatment and was operated on while the military stood guard. Join Dr. Roger Leir and follow in his footsteps as he takes you through the entire investigative journey. This could well be the most important UFO story to ever come out, and is a must read for anyone interested in the subject. **152 pages • $14.95**

Available from The Book Tree 1-800-700-TREE • secure automated voice system, all cards accepted • or send check/MO to: Book Tree, PO Box 16476, San Diego, CA 92176 • add 8% tax if in CA, $5 shipping up to 2 items media mail or $6.50 priority, will ship asap with free catalog.

CPSIA information can be obtained at www.ICGtesting.com
Printed in the USA
LVOW12s2004201014

409623LV00001B/328/P